PROJECT CONTRIBUTORS

FEDERAL PARTNERS

Lee Sobel, EPA Office of Sustainable Communities, 1200 Pennsylvania Avenue, NW, Mail Code: 1807T, Washington DC 20460, Phone: 202.566.2851, Email: sobel.lee@epa.gov

Melissa Kramer, EPA Office of Sustainable Communities

Anne Keller, EPA Region 4
Ramona McConney, EPA Region 4
Heinz Mueller, EPA Region 4
Alan Powell, EPA Region 4
Christopher Choi, EPA Region 5
Stephanie Cwik, EPA Region 5
Jon Grosshans, EPA Region 5
Deborah Orr, EPA Region 5
James Vanderkloot, EPA Region 5
Cynthia Cody, EPA Region 8

Elaine Lai, EPA Region 8
Brian Gillen, HUD
Pauline Zvonkovic, HUD
Charles Goodman, DOT
Michael Patella, DOT
Dwayne Weeks, DOT
Robert Buckley, Federal Transit Administration
Keith Melton, Federal Transit Administration
Jeff Price, Federal Transit Administration

CONSULTANT TEAM

Michael Matichich, CH2MHill
Dena Belzer, Strategic Economics
Sarah Graham, Strategic Economics
Alison Nemirow, Strategic Economics

Eli Popuch, Strategic Economics
Ellen Greenberg, Arup
Abigail Rolon, Arup

COMMUNITY TEAM

Malaika Rivers, Cumberland Community Improvement District
Brantley Day, Cumberland Community Improvement District
Tom Boland, Cumberland Community Improvement District
Reggie Greenwood, South Suburban Mayors and Managers Association
Janice Morrissy, South Suburban Mayors and Managers Association
Dave Chandler, Center for Neighborhood Technology
Dominic Tocci, Metropolitan Planning Council

Christina Oliver, Utah Transit Authority
Kathy Olson, Utah Transit Authority
Matthew Sibul, Utah Transit Authority
Nick Duerksen, Sandy City
Kasey Dunlavy, Sandy City
Jill Wilkerson-Smith, Redevelopment Agency of Salt Lake City
Gabriel Epperson, Envision Utah
Kevin Fayles, Envision Utah
Christina Oostema, Envision Utah
Kenneth Johnstone, City of Wheat Ridge
Sarah Showalter, City of Wheat Ridge

Cover photos and credits: The Crossings, Gresham, Oregon (Metro, Portland, Oregon); Seattle South Lake Union Streetcar (Steve Morgan via Wikipedia); Denver Union Station (K_Gradinger via Flickr.com). Back cover photos and credits: Corvallis, Oregon (Wendell Ward via Flickr.com); Sacramento Regional Transit District light rail (Ian Britton via FreeFoto.com); White Flint Metrorail station (EPA).

TABLE OF CONTENTS

FREQUENTLY USED ACRONYMS AND ABBREVIATIONS

CDBG	Community Development Block Grant
CMAQ	Congestion Mitigation and Air Quality Improvement
EPA	U.S. Environmental Protection Agency
FTA	Federal Transit Administration
HUD	U.S. Department of Housing and Urban Development
MPO	metropolitan planning organization
MTC	Metropolitan Transportation Commission
P3	public-private partnerships
RTD	Regional Transportation District
SAFETEA-LU	Safe, Accountable, Flexible, Efficient Transportation Equity Act: A Legacy for Users
SGIA	Smart Growth Implementation Assistance
STP	Surface Transportation Program
TIF	tax increment financing
TIFIA	Transportation Infrastructure Finance and Innovation Act
TOAH	Transit-Oriented Affordable Housing
TOD	transit-oriented development
DOT	U.S. Department of Transportation

Metropolitan areas form economic regions that benefit from passenger rail systems. Communities have learned that the benefits of public transport can be enhanced when station-area planning makes it easier for people to walk or bike as well as take transit or drive, provides affordable housing options, and offers businesses greater access to potential employees and customers from across the region. This type of planning, known as transit-oriented development (TOD), brings together housing, transportation, and jobs.

But while transit and TOD can offer a community a host of advantages, the infrastructure is costly. A street network is required to get people to their local destinations. This street network must also have infrastructure and facilities to support drivers, transit users, bikes, and pedestrians. Sidewalks and on-street parking will be needed, and commuters, residents, and commercial users often need parking garages. Energy, water, and stormwater must be addressed and managed. Regardless of who delivers the infrastructure, it must be funded, and a municipal commitment might be needed to instill market confidence.

Rail projects and TOD are long-term economic commitments. Whether a particular market is expanding or contracting, passenger rail and TOD can catalyze economic prosperity. A municipality does not want to pass up long-term transportation investments for lack of funding or financing. In many cases, places with or considering passenger rail already have professional staff with experience in sophisticated financial transactions for various types of infrastructure and transportation finance. Yet funding might already be allocated to other projects, or existing sources of funding such as revenue, formula funds, or grants might no longer be available at past levels. This raises the troubling issue of how to balance investments for long-term growth and development when the ability to fund these projects is limited.

This report provides information about funding mechanisms and strategies that communities can use to provide innovative financing options for TOD. It explains dozens of tools that provide traditional financing as well as new tools. The tools are broadly categorized under:

- **Direct fees**, including user and utility fees and congestion pricing.
- **Debt tools**, including private debt, bond financing, and federal and state infrastructure debt mechanisms.
- **Credit assistance**, including federal and state credit assistance tools and the Transportation Infrastructure Finance and Innovation Act (TIFIA).
- **Equity**, including public-private partnerships and infrastructure investment funds.
- **Value capture**, including developer fees and exactions, special districts, tax increment financing, and joint development.
- **Grants and other philanthropic sources**, including federal transportation and community and economic development grants and foundation grants and investments.
- **Emerging tools**, including structured funds, land banks, redfields to greenfields, and a national infrastructure bank.

This report also describes how 11 communities across the country have used these tools as stand-alone devices, in combination with other tools, or in phasing strategies in four categories:

- Station and station-area infrastructure financing strategies.
- District and downtown infrastructure financing strategies.
- Transit corridor infrastructure financing strategies.
- Regional initiatives.

The report also introduces four innovative models that communities could consider as they develop plans for financing infrastructure and creating TOD:

- **Anchor institution partnerships** with nonprofit or private entities such as universities, hospitals, and corporations that are inextricably tied to their locations because of real estate holdings, capital investment, history, or mission.

- **Corridor-level parking management** that would set parking prices and manage parking demand across a transit corridor or system, including both transit station parking and surrounding on- and off-street spaces.

- **Land banking** that can make it easier and more affordable to assemble and acquire land for TOD infrastructure.

- **District energy systems** that could reduce individual buildings' energy use, encourage renewable energy, and facilitate compact development.

A community's context, needs, and resources will determine which strategy or combination of strategies is most appropriate for funding TOD infrastructure. Strong markets will have more tools at their disposal than weaker markets. Certain infrastructure components such as structured parking might always be difficult to finance, whether due to costs and risk, market synergies, or project dynamics. Some communities might find that the tools are helpful but that they must overcome administrative challenges such as statutory requirements, hiring new staff, or creating new entities that have authority to originate the funding, enter into financing agreements, and administer the funding program. Some places might face the challenge of limited local capacity, such as a lack of public understanding of the opportunity, lack of local organizations to engage and partner with, or a lack of qualified developers. As they determine how to proceed, local governments could consider some guidelines for thinking strategically about TOD infrastructure:

- Have a plan that establishes a broad, long-term vision for a TOD area yet is flexible enough to respond to a changing market cycle, funding opportunities, and other conditions. Constant monitoring and proactive coordination can allow local governments to take advantage of new opportunities as they emerge.

- Think strategically about prioritizing public investments and public funds. Starting with small steps and moving forward incrementally helps to build market confidence and attract other sources of capital.

- Look for multiple funding sources.

- Look for a broad funding base, both to generate the most funding possible and to create a more stable revenue stream, which could allow the project to get a lower interest rate.

- Look for synergies among infrastructure projects. By grouping projects together, communities might be able to create efficiencies.

- Look for partnerships to fill the gaps left by traditional funding sources.

I. INTRODUCTION

This report on funding and financing for transit-oriented development (TOD) infrastructure was produced by the U.S. Environmental Protection Agency (EPA) Office of Sustainable Communities as part of the Smart Growth Implementation Assistance (SGIA) program, which helps state, local, regional, and tribal governments that need tools, resources, and other assistance to achieve their growth- and development-related goals (see EPA Smart Growth Implementation Assistance Program for more information). This report was developed by working with four communities that requested assistance from EPA on funding and financing infrastructure to support TOD: Cobb County and the Cumberland Community Improvement District, Georgia; South Suburban Mayors and Managers Association, Illinois; Utah Transit Authority, Salt Lake City and Sandy City, Utah; and city of Wheat Ridge, Colorado. The sites identified for TOD in each of the communities had different assets and challenges. However, the issues they were confronting had many commonalities that suggested a single project could help meet their needs and the needs of many communities across the country that are considering options for funding and financing infrastructure to support TOD. EPA conducted site visits at each location in the fall of 2010 and hired contractors to help develop this document.

TOD is development located within a quarter- to half-mile radius of a transit station that offers a mix of housing, employment, shopping, and transportation choices within a neighborhood or business district.[1] This easy access to public transit can lower household costs by giving people less expensive alternatives to driving, and it can give people access to more job opportunities throughout the region.

TOD often requires significant investments in infrastructure and community facilities for the type of development that can support robust transit use. These investments might include:

- Increasing the capacity of utilities (e.g., sewer, water, storm drain) and roads to support more development.

- Facilitating walking and bicycling by adding or improving sidewalks, crosswalks, bicycle lanes, bicycle storage, and streetscape enhancements such as lighting, street trees, and benches.

- Creating or improving parks, plazas, and other open space.

- Building structured parking garages for park-and-drive transit riders, which allows surface parking lots to be redeveloped for TOD.

These types of TOD infrastructure and the development they facilitate can benefit the environment, the economy, and public health by making it easy for people to walk, bicycle, or take transit; reducing pollution from automobiles; and providing affordable transportation options. However, communities often struggle to pay for TOD infrastructure because it requires upfront investment and—because many of the benefits accrue to the public and are therefore difficult to monetize—rarely generates sufficient revenue to pay for itself.

---

[1] In this report, "TOD" generally refers to an entire neighborhood or district rather than to an individual development project.

This report provides local governments with a comprehensive overview of existing tools and strategies and explores emerging, innovative models for funding and financing TOD infrastructure. While the report focuses on infrastructure such as roads, bike and pedestrian improvements, parks, streetscape improvements, structured parking, and utilities (including sewer, water, and storm drains), some of the tools and strategies can apply to other investments that are necessary to support sustainable, equitable TOD, such as providing affordable housing, acquiring and assembling land parcels, and building transit stations.

A. TRANSIT-ORIENTED DEVELOPMENT AND SMART GROWTH

Development decisions have direct and indirect consequences for the built and natural environments and public health. Where and how we build directly affects wildlife habitat and water quality by replacing natural cover with impervious surfaces like asphalt and concrete. Development patterns that separate land uses and neighborhoods that provide few transportation options foster reliance on the automobile, contributing to greenhouse gas emissions that cause global climate change and other pollution that harms air quality and causes other environmental and public health problems.

Smart growth practices can lessen the environmental and health impacts of development by building compactly and mixing land uses, which can make walking, bicycling, and transit use more appealing by putting destinations closer together. Compact development can reduce impervious surfaces, which protects water quality by reducing the amount of polluted runoff that flows into surface waters. Using land more efficiently takes development pressure off of environmentally sensitive areas. Smart growth strategies encourage a mix of housing types at different price points to allow people at all stages of life to live in the same neighborhood. Encouraging investment in existing communities takes advantage of previous investments, using public funds more efficiently.

TOD districts feature compact, multistory development that uses land and other resources more efficiently; a mix of residential and commercial uses; and streets designed to make walking, biking, and transit safe and practical. TOD can generally be built at greater densities because it is close to transit. If it relied strictly on the road network for transportation, such densities could cause major traffic congestion. TOD takes many different forms, with different land uses and building densities, depending on the context of the station area.

B. THE BENEFITS OF TOD

TOD makes it easier for those who live or work in the area around the station to get around the region. It also benefits drivers because it removes trips from the road network. The mix of commercial and residential uses, enhanced pedestrian realm and streetscapes, and reduced traffic congestion improve quality of life in transit-oriented neighborhoods. Including affordable housing in TOD offers these benefits to lower-income households who need them most; transportation expenses can be a significant proportion of household expenditures. Neighborhoods that are walkable and have access to transit and a variety of stores and services are "location efficient," and location efficiency correlates strongly with household transportation spending. Transportation costs rise from an average 15 percent of household income in location-efficient neighborhoods to an average 28 percent of income in non-location-efficient neighborhoods.[2] Enabling workers in households at all income levels to reach job centers without long, expensive commutes helps promote regional economic prosperity. Finally, moving away from development patterns that give people no choice but to drive to every destination helps reduce greenhouse gas emissions and other pollution.[3]

C. THE CHALLENGE OF FUNDING INFRASTRUCTURE FOR TOD

Transit corridors connect several transit station areas that often have very different development patterns and market strengths, ranging from downtowns and other urban districts that already have high-density residential, office, and retail development to suburban neighborhoods that typically have spread-out, single-use development. The infrastructure needs of a station area depend on its development context, as well as on factors like the capacity of existing facilities and the planned increase in development intensity. For example, in some station areas the existing street network, sidewalks, and other infrastructure are designed to serve spread-out development that requires automobile use. To achieve the benefits of TOD in these areas, the public sector and/or developers might need to provide new or improved pedestrian and bicycle facilities, roads, utilities, and public open space to link residents and workers to transit and support the increased population that higher-density development will bring. Expensive structured parking facilities might be required to accommodate park-and-ride commuters, allowing development to occur next to transit stations on land that might otherwise be occupied by surface parking lots. Even in more densely populated, urban districts, new development might require improving infrastructure capacity, while TOD in previously undeveloped "greenfield" areas might involve significant investments in new infrastructure systems and community facilities. Successful TOD can also require public-sector help with acquiring and assembling land from multiple owners, affordable housing development, transit station construction, and other activities that make TOD possible.

TOD infrastructure such as transit facilities, sidewalks, utilities, and affordable housing can provide significant public benefits, such as improving public health by reducing vehicle emissions. However, infrastructure and related investments are costly. Moreover, purely public projects like sidewalks and local roads rarely generate any revenue. Services like water, wastewater systems, and parking can generate revenue for operations and maintenance from users, but raising rates high enough to pay for significant new capital investments can be contentious and requires careful planning to secure the necessary support.[4] To add to the funding challenge, TOD infrastructure and community facilities often

---

[2] Haas, Peter M.; Makarewicz, Carrie; Benedict, Albert; and Bernstein, Scott. "Estimating Transportation Costs by Characteristics of Neighborhood and Household." *Transportation Research Record* 2077:62-70. 2008.

[3] A 2010 study of the Chicago Metropolitan Region indicates that by living within half a mile of transit, the average household reduces its transportation-related greenhouse gas emissions by 43 percent. Households living near the most location-efficient central city transit zones reduce their emissions by 78 percent. Center for Transit-Oriented Development and Center for Neighborhood Technology. *Transit Oriented Development and the Potential for VMT-related Greenhouse Gas Emissions Growth Reduction.* March 2010. http://www.reconnectingamerica.org/resource-center/browse-research/2010/transit-oriented-development-and-the-potential-for-vmt-related-greenhouse-gas-emissions-growth-reduction.

[4] For example, *Avoiding Rate Shock: Making the Case for Water Rates*, a 2004 publication from the American Water Works Association, documents the challenges of securing support for water rate increases and offers strategies to improve the chances of success by connecting financial and rate planning technical studies with stakeholder outreach and additional steps.

need to be in place *before* new private development can occur—either because additional infrastructure is required to support new uses, or, in a place with a weak real estate market, to make a location attractive for developers, residents, and workers.

Providing TOD infrastructure is further complicated by the number of entities that can be involved. Local governments have typically provided local roads, bicycle and pedestrian facilities, open space, utilities, and public parking, although many localities are shifting some of this responsibility to developers. Transit agencies also play an important role by building and maintaining transit stations, parking, and bicycle and pedestrian facilities and sometimes by forming partnerships to develop agency-owned land. Regional transportation planning organizations, states, and the federal government also play a role, typically by funding and financing infrastructure and setting the rules that govern the use of those funds.

The challenges of funding and financing TOD infrastructure call for continued innovation and creativity in identifying appropriate funding and financing tools and combining those tools into comprehensive strategies. This report reviews the tools and strategies that local governments, in partnership with transit agencies and regional, state, and federal government can deploy to meet the challenges of paying for the infrastructure required to attract and support TOD. Some of these tools and strategies have rarely been applied to TOD infrastructure and might require modification to apply in TOD contexts. The descriptions and examples in this report are intended to help local governments learn about these tools, encourage consideration of these emerging approaches, and, where appropriate, spur the development of modified tools.

II. OVERVIEW OF TOOLS AND STRATEGIES FOR FUNDING AND FINANCING TOD INFRASTRUCTURE

Communities can fund and finance TOD infrastructure using many tools, strategies, and innovative models. This overview provides context for understanding their detailed descriptions that follow. The overview includes:

- Background information, including definitions of terminology and a discussion of the roles of different governmental and nongovernmental entities in providing TOD infrastructure.

- A description of common and emerging tools for funding and financing TOD infrastructure.

A. BACKGROUND ON INFRASTRUCTURE FUNDING AND FINANCING

TERMINOLOGY

The first step in paying for infrastructure is identifying a funding source. In the context of infrastructure development, a funding mechanism refers to a revenue stream or source. Some types of infrastructure generate revenue directly by charging users a fee. For example, transit systems, many parking facilities, water and wastewater systems, and toll roads and bridges charge user fees, which can be used for either operations and maintenance or capital improvements. Other types of infrastructure, like sidewalks, bike racks, local roads, and parks, rarely generate revenue directly because they are free to use. To pay for this type of non-revenue-generating infrastructure, local governments typically rely on revenue from taxes, fees, and other sources.

Once a funding source is identified, local governments can approach paying for infrastructure in two ways: pay-as-you-go or financing. In a pay-as-you-go approach, an improvement is made only once enough revenue has been collected to cover the cost of the improvement. By contrast, with a financing approach, the improvement is paid for before revenue equal to the full cost of the improvement is available, typically by borrowing against future revenue and issuing bonds that are paid back over time with taxes, user fee payments, or other revenue sources.

In this report, tools are funding and financing sources that local governments or transit agencies can use to pay for specific types of infrastructure. Strategies are action plans that public agencies create and implement to achieve a goal, such as attracting new development or promoting walking, bicycling, and transit use in a station area. A TOD infrastructure financing strategy typically includes such elements as:

- A clear vision and goals for a particular geographic area. (The strategies discussed in this report largely concern entire neighborhoods or districts rather than individual development projects.)

- An assessment of the local real estate market context.

- A list of key infrastructure needs and associated costs.

- A phasing plan that considers which infrastructure improvements are required and in what order to support planned development.

- A discussion of which public agencies and private entities will have a role in implementation and which entity will take the lead on implementing each project.

- An assessment of potential funding and financing sources tailored to the infrastructure needs, market conditions, and capabilities of the implementing entities. In some cases, funding source availability can be as important as the infrastructure phasing assessment in determining which infrastructure projects get financed and in what sequence.

The case studies in the report illustrate the various components of TOD infrastructure financing strategies and examples of how local governments and their regional and state partners have used funding and financing sources in new ways, often with private and nonprofit partners. The word "model" is used in the report to refer to innovative approaches for funding and financing infrastructure.

ROLES IN FUNDING AND FINANCING TOD INFRASTRUCTURE

The roles of different entities in infrastructure provision vary depending on the state and jurisdiction. In general, the main players in funding and financing TOD infrastructure are:

- Local governments: Cities, towns, counties, and other local government entities have traditionally been responsible for building and maintaining basic local infrastructure like sewer, water, other utilities, roads, bicycle and pedestrian improvements, and public parking. In some cases, local governments have established special districts or municipal utilities to operate revenue-generating infrastructure such as a sewer or water system. Local governments sometimes also rely on partnerships with private entities to deliver infrastructure, although they typically retain the primary responsibility for non-revenue-generating infrastructure (e.g., parks and sidewalks).

- Transit agencies: In most places, a specially constituted agency or authority, often with its own revenue stream in the form of a local sales tax or other levy, is charged with building, owning, and operating transit facilities, including rail lines, buses, transit stations, and station parking lots or structures. In addition to being involved in providing station area infrastructure, transit agencies can work directly on TOD when they have property to develop.

- Metropolitan planning organizations (MPOs): MPOs are federally mandated organizations charged with planning for transportation improvements and distributing federal transportation dollars in urbanized areas. In some states, MPOs are also responsible for allocating state transportation dollars in their regions. Some of the federal money MPOs receive is flexible and can be used to pay for many components of TOD infrastructure.

- State government: Most states have a limited role in developing and maintaining the types of local infrastructure discussed in this report. However, states play a significant role in distributing federal funding for infrastructure, particularly in rural regions that do not have MPOs. In addition, many states have established their own funding and financing programs for infrastructure (typically using tax revenue and bonds), and state legislatures largely determine the types of tools that local governments have at their disposal. For example, state statutes define whether and how local governments can establish tax-increment financing districts, special assessments, and other types of taxing and debt mechanisms.

- Federal government: The federal government plays a critical role in funding transportation, water and sewer systems, green space, and other types of infrastructure, as well as environmental protection and cleanup, housing, community and economic development, and other related activities. Much of the funding for transportation, housing, and community and economic development is distributed in the form of block grants to states, MPOs, or local governments, which have significant discretion in allocating funds. Federal agencies also provide technical assistance, conduct research, and help share knowledge across the country.

B. TOOLS FOR FUNDING AND FINANCING TOD INFRASTRUCTURE

In a time of severe fiscal constraints for many public entities, communities are looking for ways to make the best use of local government revenue, such as property and sales taxes, and generate new revenue to fund TOD infrastructure. A key task in creating a TOD infrastructure financing strategy is to evaluate which tools will work best for a particular project or in a particular development context. Beyond general property and sales taxes, the tools that local governments and transit agencies use to fund and finance TOD and other infrastructure fall into six broad categories:

- Direct fees.

- Debt.

- Credit assistance.

- Equity.

- Value capture.

- Grants and other philanthropic sources.

In addition, there are some emerging tools that do not fit neatly into one of the previous six categories or are new concepts still being developed.

The description of each tool includes the types of places where it could be most useful. For example, some tools depend on a strong real estate market and property value appreciation to generate revenue, while others are well-suited to weaker-market areas. Some tools are available only where the state legislature has authorized them.

Several of the tools can be used only for projects that meet minimum cost thresholds (these thresholds are noted where they apply). In general, however, few rules of thumb apply for determining how large or small a project must be to use a tool. Instead, communities must consider whether a project is of sufficient size to justify the transaction costs of accessing a funding source. Depending on the tool, those costs could include writing a grant application or structuring a complex financial transaction. Regional and local priorities will also determine whether a tool is applicable to a specific project. This is especially true of federal funding sources, many of which are distributed as block grants that allow state, regional, or local governments significant discretion over allocation. For example, all MPOs receive Congestion Mitigation and Air Quality Improvement (CMAQ) Program and Surface Transportation Program (STP) dollars, but states vary in the degree of flexibility they allow MPOs to use in allocating those funds, and MPOs vary in the priority they put on spending that money on TOD-related improvements.

More detailed tool descriptions and case studies showing how they have been used are presented in Appendix B and Chapter III, respectively.

DIRECT FEES

Direct fees charge people for using public infrastructure or goods. Fees can be charged for new and existing development and are therefore applicable in strong and weak real estate markets. However, the rate at which a fee can be set generally depends on local conditions; for example, parking fees or bridge tolls can be set higher in places with strong demand from drivers.

- **User fees and transportation utility fees:** User fees and rates are charged for the use of public infrastructure or goods, including transit, parking facilities, water or wastewater systems, and toll roads or bridges. Local governments or utilities might be able to issue bonds backed by user fee revenue to pay for new or improved infrastructure. Such fees and rates are typically set to cover a system's yearly operating and capital expenses, including annual debt service for improvements to the system.

- **Congestion pricing:** Congestion pricing manages demand for services by adjusting prices depending on the time of day or level of use. Congestion pricing has been used to mitigate traffic congestion, with revenue used to cover costs, support transit service, or improve the highway system.

DEBT

Debt tools are mechanisms for borrowing money to finance infrastructure. Local governments can access credit through private financial institutions, the bond market, or other, specialized mechanisms that the federal government and states have established for financing particular types of infrastructure. Local governments can issue debt for projects that do not generate revenue (typically in the form of general obligation bonds), but most types of debt must be secured by revenue generated either by the infrastructure that the debt is used to fund (e.g., parking or utility fees) or within the geographic area that will benefit from the improvement (e.g., tax-increment financing generated by property or sales tax increases can typically be used to pay for improvements only in a specified area). Except for debt that is secured by revenue such as property taxes that is related to real estate performance, the availability of debt is not usually related directly to the strength of the local real estate market. Rather, potential lenders, including private financial institutions as well as bond investors, decide how much they are willing to lend and on what terms based on the creditworthiness of the borrower and the reliability of the revenue stream that will be used for repayment (e.g., taxes, user fees, or leases).

- Private debt: Public entities can borrow money from commercial banks, **industrial loan companies or industrial banks** (banks owned by a non-financial corporation), and other private financial institutions to finance revenue-generating infrastructure. However, publicly issued debt (i.e., bonds) is typically less costly.

- Bond financing: Because most publicly issued bonds are exempt from state and federal taxes, public entities can typically access lower interest rates by issuing bonds rather than by borrowing money from a private lender. The most common types of bonds include:

  - **General obligation bonds:** General obligation bonds are backed by the "full faith and credit" of the issuer rather than the revenue from a specific project and can therefore be used to finance infrastructure that does not generate revenue. General obligation bonds are tax-exempt and can be issued by governmental entities at the state or local level, including counties, cities, transit agencies, special-purpose districts, public utilities, and school districts.

- o **Revenue bonds:** Revenue bonds are issued for municipal projects that generate revenue and are secured by (i.e., repaid solely by) the revenue generated by the facility they finance (e.g., farebox revenue from a transit system, user fees from a parking garage or utility, or tolls from a road or bridge). Like general obligation bonds, revenue bonds are tax-exempt and can be issued by governmental entities at the state or local level.

- o **Private activity bonds:** Private activity bonds are issued by state or local governments (and are therefore exempt from state and federal taxes) and apply the proceeds of the bonds to private business purposes that have a public benefit.[5] Like revenue bonds, private activity bonds are secured by and paid from the revenue of the project for which the bonds are sold.

- o **Certificates of participation and lease revenue bonds:** Certificates of participation and lease revenue bonds are tax-exempt bonds issued by state or local governments that are secured with revenue from the lease of land, public infrastructure, and transportation assets, including parking structures, rail transit, water and wastewater treatment plants, and other public facilities.[6]

- Specialized debt for infrastructure: In addition to the bonds described above, local governments can sometimes access debt mechanisms designed by the federal government or states to finance particular types of infrastructure. In some cases, these debt mechanisms could not be used directly for TOD infrastructure as defined in this report but could help make TOD infrastructure projects possible by funding transit or roads, freeing up funds for other uses. Examples of these debt mechanisms include:

  - o **Revolving loan funds:** A revolving loan fund is a pool of money dedicated to specific kinds of investments. The money used to repay loans replenishes the fund and is loaned out again. Initial funding sources for revolving loan funds are typically public or private seed money, such as a grant, other public funds, or the one-time proceeds from sale of an asset, and/or an ongoing revenue stream such as a dedicated portion of a new or existing tax. Revolving loan funds can provide low-interest loans and access to capital markets for projects that would otherwise have difficulty securing financing if they meet economic development, environmental, or other public policy goals. In contrast to a structured fund (discussed below), which is capitalized by investors with an expectation of return, as borrowers repay their loans, the money can be lent again to new borrowers and revolve indefinitely.

  - o **State infrastructure banks:** Many states have established state infrastructure banks, which provide local governments with low-interest loans. State infrastructure banks typically function as revolving loan funds.

  - o **Grant anticipation revenue vehicle (GARVEE) bonds:** GARVEE bonds are federally tax-exempt debt mechanisms backed by federal appropriations for transportation projects that are not expected to generate revenue. Most commonly used for highway construction, GARVEE bonds

---

[5] FHWA. Private Activity Bonds (PABs). http://www.fhwa.dot.gov/ipd/fact_sheets/pabs.htm. Accessed August 23, 2012.

[6] AASHTO Center for Excellence in Project Finance. Certificates of Participation. http://www.transportation-finance.org/funding_financing/financing/bonding_debt_instruments/certificates_of_participation.aspx. Accessed August 24, 2012.

can also be used for transit and other transportation projects funded by other federal grant programs such as the STP and the CMAQ Program, described below. Local governments must work with MPOs and state departments of transportation to access GARVEE bonds, which also must be approved by the U.S. Department of Transportation (DOT).

o **Railroad Rehabilitation and Improvement Financing (RRIF):** The RRIF program, administered by DOT, provides loans and loan guarantees that can be used to acquire, improve, or rehabilitate rail facilities or facilities that connect rail to other forms of transportation. Eligible RRIF borrowers include railroads and state and local governments.[7]

CREDIT ASSISTANCE

Credit assistance improves a borrower's creditworthiness by providing a mechanism that reduces the chances of a default. Borrowers can thus access better borrowing terms, which can expedite the implementation of infrastructure projects. Credit assistance tools require some source of revenue to pay back debt; their use is not otherwise linked to the strength of the local real estate market.

- **Credit assistance tools:** Federal and state agencies have developed a variety of financial tools to help local governments access credit to expedite infrastructure projects. Credit assistance improves local agencies' creditworthiness and thus lets them access better borrowing terms and lower financing costs. Credit assistance can take many forms including bond insurance, credit enhancements, credit lines, loans, and loan guarantees.

- **Transportation Infrastructure Finance and Innovation Act (TIFIA):** TIFIA is a DOT-administered program that provides federal credit assistance to state and local government entities for large (with total project costs of $50 million or more) surface transportation projects, such as transit projects and highways, that have dedicated funding sources.[8] As with some debt mechanisms, TIFIA might not apply to TOD infrastructure as defined in this report, but the program could help make TOD infrastructure projects possible by funding transit or roads and freeing funds for other uses.

EQUITY

Equity tools allow private entities to invest (i.e., take an ownership stake) in infrastructure in expectation of a return. Unless the public sector is willing to directly pay the private partner for constructing, financing, operating, and/or maintaining a facility, equity sources are typically available only for infrastructure that generates a significant return, such as parking facilities, utilities, toll roads, or airports. The availability of equity is not typically tied to the strength of the local real estate market, except insofar as the potential source of revenue is tied to real estate values.

- **Public-private partnerships:** A public-private partnership is a contractual agreement between a public agency and a private-sector entity whereby "the skills and assets of each sector (public and private) are shared in delivering a service or facility for the use of the general public. In addition to

---

[7] Federal Railroad Administration. "Railroad Rehabilitation & Improvement Financing (RRIF) Program." http://www.fra.dot.gov/rpd/freight/1770.shtml. Accessed August 22, 2011.

[8] de la Pena, Patricia; Caplicki, Edmund V.; and Santiago, Simon J.. "2010 Transportation Infrastructure Year in Review." Nossaman LLP. February 17, 2011. http://www.nossaman.com/7749.

the sharing of resources, each party shares in the risks and rewards in the delivery of the service and/or facility."[9] In a typical public-private partnership, the private entity provides the capital cost to finance a public project, such as a parking facility, toll road, or airport, then collects some portion of the revenue generated by the project. In most public-private partnerships, the public-sector partner guarantees payment to the private-sector partner even if the project does not deliver the expected level of revenue or if the expected revenue does not cover the entire cost of debt repayment.

- **Infrastructure investment funds:** Infrastructure investment funds are pools of funds collected from many investors for the purpose of investing in infrastructure, often in the form of a public-private partnership. These funds are typically repaid through user fees.

VALUE CAPTURE

Value capture tools capture a portion of the increased value or savings resulting from publicly funded infrastructure. Value capture mechanisms are typically established by a local government in accordance with state law. They sometimes require a vote by the affected property owners. Depending on the tool, value capture can entail the creation of a new assessment, tax, or fee (e.g., a special tax or development impact fee); the diversion of new revenue generated by an existing tax (e.g., tax-increment financing); or a revenue-sharing agreement that allows a government agency to share some of the revenue generated by developing publicly owned land (e.g., joint development). Value capture tools are generally most applicable to strong real estate markets because they depend to some extent on new development or property value appreciation to generate revenue.[10]

Depending on the predictability of the revenue stream, value capture mechanisms can either be used for pay-as-you-go improvements or, when the revenue stream is expected to be consistent over time, as with a special assessment or tax-increment financing, can finance the issuance of revenue bonds. Although state law usually defines how and where these mechanisms can be used, they are typically not confined to revenue-generating infrastructure and can be used to fund all types of TOD infrastructure, including utilities, roads, pedestrian and bicycle improvements, and parking facilities.

- **Developer fees and exactions (impact fees, system development charges, facility fees, infrastructure reimbursement agreements, developer exactions):** Development impact fees and exactions are charges assessed on new development to defray the cost to the jurisdiction of expanding and extending public services to the development.[11] The fees are generally collected once and are used on a pay-as-you-go basis to offset the cost of providing public infrastructure such as new streets and utilities. Because these are one-time fees, they cannot be used for ongoing facility operations and maintenance.

- **Special districts (benefit assessment districts, business improvement districts):** Special districts are formed to include a geographical area in which property owners or businesses agree to pay an

---

[9] National Council for Public-Private Partnerships. "How PPPs Work." http://www.ncppp.org/howpart/index.shtml#define. Accessed August 2011.

[10] Center for Transit-Oriented Development. *Capturing the Value of Transit*. Prepared for FTA. 2008. http://reconnectingamerica.org/resource-center/books-and-reports/2008/capturing-the-value-of-transit-3.

[11] Ibid.

assessment to fund a proposed improvement or service from which they expect to benefit directly. Special districts can be used either for pay-as-you-go improvements or to finance the issuance of bonds backed by the assessment revenue. They can be used for a wide range of projects, including pedestrian, bicycle, and streetscape improvements and utilities.[12] Depending on the state enabling legislation under which a special district is formed, assessment districts might be able to pay both capital expenditures and operations and maintenance costs.

- <u>Tax increment financing (TIF)</u>: TIF works differently according to laws in each state but typically captures the increase in property tax revenue (and, in some states, sales tax revenue) that occurs in a designated area after a set year. The tax increment is collected for a set period (usually between 15 and 30 years) and the tax increment can be used to secure a bond, allowing the issuer to collect the money up front, or it can be used on a pay-as-you-go basis over time. The most common uses of TIF are for local infrastructure, environmental cleanup, and land assembly.[13]

- <u>Joint development</u>: Joint development is the only value capture mechanism transit agencies commonly use. It is generally a real estate development project that involves coordination among multiple parties to develop sites near transit, usually on publicly owned land, and can take many forms, ranging from an agreement to develop land owned by the transit agency to joint financing and development of a project that incorporates both public facilities (e.g., parking garages) and private development. Typically the transit agency and the private developer will agree to share costs of and revenue from the project.[14]

GRANTS AND OTHER PHILANTHROPIC SOURCES

Grants are funds that do not need to be paid back and are typically provided by a higher level of government to a lower level of government (e.g., from the federal government to states or localities, or from states to local governments) or by a philanthropic entity. This report discusses the federal grants that are commonly applied to TOD projects, including transportation and community and economic development grants, as well as some of the most common philanthropic investments in TOD. Most states also provide their own grant opportunities that can be used for TOD infrastructure. With the exception of grants that focus on addressing poverty or other conditions related to weak markets, grants do not usually depend on local market strength.

- Federal transportation grants: Local governments typically access these federal transportation funds through MPOs and/or state departments of transportation. Federal grants that can be used for TOD infrastructure include:[15]

  o <u>Congestion Mitigation and Air Quality Improvement (CMAQ) Program</u>: This program funds transportation projects or programs that contribute to improving air quality and relieving congestion, including pedestrian and bicycle improvements, transit, and demand management

---

[12] Ibid.

[13] Ibid.

[14] Ibid.

[15] The latest transportation bill set aside funds for a new pilot program for TOD planning, including increasing pedestrian access and enabling mixed use. As of October 2012, the details of the program have not been developed.

projects that support better decision-making for travelers choosing modes, times, routes, and locations.

- o **Transportation Alternatives Program:** This program funds a wide range of TOD infrastructure projects, including pedestrian and bicycle access improvements, streetscape improvements.

- o **Urbanized Area Formula Funding Program:** The federal Urbanized Area Formula Funding Program funds transit capital costs, maintenance of passenger facilities, and transportation-related planning.[16]

- **Federal community and economic development grants:** The federal government has several grant programs dedicated to housing for low-income households and other community and economic development. While these tools are not focused on TOD infrastructure, they can be used as part of a larger TOD project.

  - o **Community Development Block Grant (CDBG) Program:** The CDBG Program, administered by the U.S. Department of Housing and Urban Development (HUD), is intended to ensure decent affordable housing, community services to vulnerable neighborhoods, and job creation and retention of businesses. CDBG provides annual formula grants to local government agencies and states in several program areas.[17] This tool is not focused on TOD infrastructure but could be used in combination with other funding and financing tools for a larger TOD project that meets CDBG criteria.

  - o **Economic Development Administration (EDA) Grants:** EDA, an agency in the U.S. Department of Commerce, provides grants to economically distressed communities to generate new employment, help retain existing jobs, and stimulate industrial and commercial growth. Some EDA funding is reserved for public works projects, which can include a wide range of infrastructure types provided the project has an economic development purpose. Local governments apply directly to the EDA when grants are available.

- **Philanthropic sources.**

  - o **Foundation grants:** Foundations, including private foundations and public charities, are nongovernmental organizations that make grants with a charitable purpose. Studies[18] have found that foundations are interested in supporting TOD. Most of their funding to date has provided affordable housing or social services around transit facilities or even funded the transit itself. However, they may also be open to funding the infrastructure to support TOD.

---

[16] FTA. "Chapter 53 of title 49, United States Code, as amended by MAP-21." 2012.

[17] HUD. "Community Development Block Grant Program—CDBG." http://portal.hud.gov/hudportal/HUD?src=/program_offices/comm_planning/communitydevelopment/programs. Accessed August 31, 2011.

[18] Katherine Pease & Associates. *Convening on Transit Oriented Development: The Foundation Perspective.* Prepared for Center for Transit-Oriented Development, Living Cities, and Boston College Institute for Responsible Investments. February 2009. http://www.katherinepease.com/Convening%20on%20TOD%20-%20The%20Foundation%20Perspective.pdf.

o **Program-related investments:** Foundations make program-related investments to support their philanthropic mission and leverage their donations. Unlike grants, foundations expect program-related investments to be repaid, although production of income or appreciation of property cannot be a significant purpose. Program-related investments allow the recipient to borrow capital at lower rates than might otherwise be available. For the funder, the principal benefit is that the repayment or return of equity can be recycled for another charitable purpose, assuming the investment is repaid. While many program-related investments in the past have supported affordable housing and community development, they have also funded capital projects ranging from rehabilitating historic buildings to preserving open space and wildlife habitat.

EMERGING TOOLS

In addition to the established tools discussed above, several new concepts for making TOD infrastructure possible are emerging, including:

- **Structured funds:** A structured fund is a loan fund that pools money from different investors with varying risk and return profiles. Structured funds have a dedicated purpose, which is clearly defined before the fund is formed, and are managed by professionals with fund formation and loan underwriting experience. Communities have been increasingly interested in using structured funds as a property acquisition tool to support affordable housing development, particularly near transit. Structured funds are discussed in greater detail in both Appendix B, Section G-1APPENDIX C and Fundamentals of Structured Funds

- **Land banks:** Land assembly and acquisition can be a challenge for TOD because land near transit is often scarce and generally costs more. Land banks are not funding or financing sources, but communities' interest in their applicability to TOD has been growing because they are used to acquire property and are often linked to a social mission, such as neighborhood stabilization or affordable housing. While land banks have not been used for TOD infrastructure, assembling developable land in station areas could make TOD and the associated infrastructure projects more feasible.

- **Redfields to greenfields:** Redfields to greenfields is a concept for converting underused or distressed properties into an asset. A local government agency acquires underused properties (redfields) in an area and converts them into new parks (greenfields). Redfields to greenfields is not tied to any particular funding source; in fact, the local government would need to identify a funding source to pay for property acquisition and convert the property into a park, which could include parks that are part of a mixed-use TOD. The new park could boost property values of surrounding properties, increasing property tax revenue.[19]

- **National infrastructure bank:** A national infrastructure bank would finance transportation and potentially other types of infrastructure across the country by providing federal credit assistance, such as direct loans and loan guarantees to local governments. The United States does not currently have such a bank, but Congress has considered several proposals that would encourage investment in infrastructure from nonfederal sources through a mostly self-sustaining entity.

---

[19] American Planning Association. *How Cities Use Parks for Economic Development.* 2002.
http://www.planning.org/cityparks/briefingpapers/economicdevelopment.htm.

III. CASE STUDIES: COMBINING FUNDING AND FINANCING SOURCES TO FORM STRATEGIES

Eleven case studies illustrate how the tools discussed in Chapter II have been used in various combinations to fund and finance TOD infrastructure. These are cutting-edge TOD projects and plans from across the country that illustrate the variety of ways in which local and regional governments and transit agencies can combine funding and financing sources to form TOD infrastructure financing strategies. The case studies are:

- Station and station-area infrastructure financing strategies.

  o Special assessment district: New York Ave-Florida Ave-Gallaudet University Metrorail Station (Washington, D.C.).

  o Joint development: West Dublin BART (Dublin, California).

  o Federal loans, grants, and credit enhancement: Denver Union Station (Denver, Colorado).

- District and downtown infrastructure financing strategies.

  o Special assessment district: Downtown Stamford (Stamford, Connecticut).

  o Public-private partnership: New Quincy Center (Quincy, Massachusetts).

  o Special assessment and density incentives: White Flint Sector Plan (Montgomery County, Maryland).

- Transit corridor infrastructure financing strategies.

  o Multistation tax-increment financing: Dallas TIF for TOD (Dallas, Texas).

  o Corridorwide tax-increment financing: Atlanta BeltLine (Atlanta, Georgia).

- Regional initiatives.

  o Supporting TOD with federal transportation dollars: Transportation for Livable Communities (San Francisco Bay Area, California).

  o Structured funds for TOD land acquisition: Bay Area Transit-Oriented Affordable Housing Acquisition Fund (San Francisco Bay Area, California).

  o Regional TOD investment framework: Central Corridor Light Rail and the Central Corridor Funders Collaborative (Twin Cities, Minnesota).

Exhibit 1 summarizes the eight local strategies (excluding the regional initiatives). They cover various geographic scales and development contexts, many types of infrastructure, and a wide range of project costs. While the combination of tools used in each case study is different, there are some commonalities. Each plan required a combination of funding and financing sources tailored to the specific project, real estate market context, and state and local laws and resources. Most of the projects include one or more value capture mechanisms (typically development fees, a special assessment district, and/or a tax-increment financing district) and multiple federal transportation grants (usually accessed through the MPO), as well as a variety of state grants and local sources such as sales or property tax revenue.

Exhibit 1 also illustrates that local governments tend to have more funding and financing tools at their disposal than do transit agencies. In particular, the only value capture mechanism available to most transit agencies is joint development. Because local governments may have access to more types of resources than transit agencies, they are responsible for the provision of most local infrastructure and have authority over land use planning and regulation. Local governments typically take the lead in planning and implementing TOD infrastructure strategies, while transit agencies focus largely on the station area infrastructure—like parking facilities, pedestrian walkways, and bicycle parking—located on land owned by the agency. Transit agencies may become more actively involved in TOD when they own land they wish to develop.

The 11 case studies illustrate the factors that local governments, transit agencies, MPOs, or other project sponsors would likely consider in forming a strategy to fund and finance TOD infrastructure, including factors such as the infrastructure required, project size, local market strength, and limitations imposed by state and local law. The case studies are generally organized around the following sections:

- Introduction and project or program background: Provides context on the project area to explain why the tools used in the case study were appropriate and describes the planning process that created the strategy.

- Funding sources and financing mechanisms: Describes how the major tools used in the example work together.

- Lessons learned: Reviews the keys to success and key barriers that the project sponsor(s) encountered in creating or implementing the strategy and that other entities attempting to use this strategy are also likely to encounter and discusses applicability of the strategy to other places.

Project and Location	Development Context	Infrastructure Types	Primary Project Sponsor(s)	Estimated Project Cost	User Fees	Bond Financing	RRIF and TIFIA	Private Financing	Developer Fees	Assessment Districts	Tax Increment	Joint Development	Federal Funds	State/Local Funds
Stations and Station Areas														
New York Ave. Station (Washington, D.C.)	Weak market; new station; infill	Transit station land acquisition and construction	Local government	$104 million						X			X	X
West Dublin BART (Dublin, CA)	Strong market; new station; greenfield	Transit station, parking, bike/ped	Transit agency	$106 million	X	X						X		
Denver Union Station (Denver, CO)	Strong market; existing station; infill	Utilities, roads, bike/ped, rail and bus stations, parking	Transit agency; local government; regional council of governments; state DOT	$435-$519 million	X	X	X	X		X	X		X	X
Districts and Downtowns														
Downtown Stamford (Stamford, CT)*	Strong market; existing station; infill	Ongoing operations and maintenance	Business improvement district; local government	N/A*						X			X	X
New Quincy Center (Quincy, MA)	Strong market; existing station; infill	Utilities, roads, bike/ped, streetscape, parking	Local government	$277 million	X		X	X		X				X
White Flint Sector Plan (Montgomery County, MD)	Strong market; existing station; infill	Roads, bike/ped, streetscape, parks, transit station, police, fire, library	Local government	$313 million		X			X	X				X
Transit Corridors														
Dallas TIF for TOD (Dallas, TX)**	Strong and weak markets; existing transit line; infill	Utilities, roads, parks, bike/ped, streetscape, affordable housing, transit improvements	Local government	N/A**		X				X	X		X	X
Atlanta BeltLine (Atlanta, GA)	Strong and weak markets; new transit line; infill	Transit, parks, bike/ped, streetscape, schools, affordable housing	Local government	$2.1-$2.7 billion		X				X	X		X	X

Exhibit 1. Summary of Case Study TOD Funding and Financing Strategies

*The Stamford special assessment district pays for ongoing operations and maintenance, not capital improvements as in the other case studies.

**The Dallas corridorwide TIF district funds individual projects rather than having a list of planned infrastructure improvements for the district.

A. SPECIAL ASSESSMENT DISTRICT: NEW YORK AVE-FLORIDA AVE-GALLAUDET UNIVERSITY METRORAIL STATION

INTRODUCTION

The New York Avenue-Florida Avenue-Gallaudet University Metrorail Station (New York Avenue Station) in Washington, D.C., constructed in November 2004, is an example of transportation infrastructure financing through a public-private partnership among local landowners, the local and federal governments, and the transit agency. It shows how government agencies can use special assessment districts to support financing and accelerate project delivery instead of using a pay-as-you-go approach. [20]

PROJECT BACKGROUND

New York Avenue Station, which is on Metrorail's Red Line, is located east of the intersection of New York and Florida Avenues and adjacent to the Amtrak Northeast Corridor railroad tracks. The station serves NoMa, an area north of Massachusetts Avenue that is near Union Station and Capitol Hill (Exhibit 2). The city had long targeted NoMa for redevelopment due to its proximity to downtown, its stagnant economy, and the large number of vacant properties in the area. By 1999, there were 5,600 people living within one-half to three-quarters of a mile of the intersection of New York and Florida Avenues. The population was 90 percent minority and had a median income of $23,296, well below the city's median income level of $30,727. An estimated 50 percent of residents did not own a car, increasing the need for transit options. Before the station opened in 2004, inadequate transportation facilities limited the attractiveness of the area for new, large-scale development.

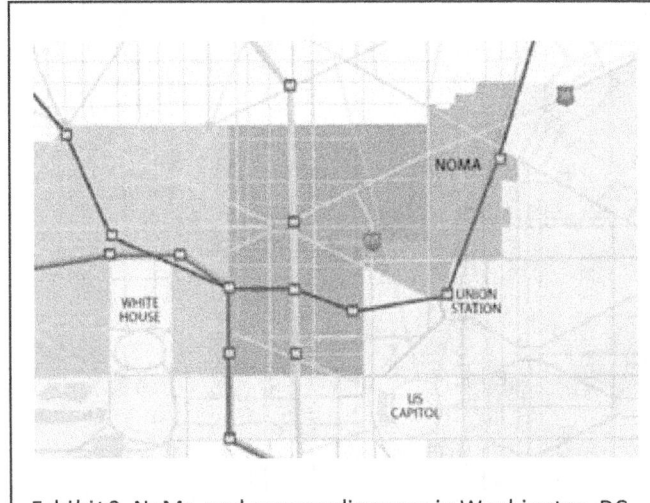

Exhibit 2. NoMa and surrounding area in Washington, DC.
Source: NoMa Business Improvement District.

In 1998, Washington, D.C., produced an economic development plan, *The Economic Resurgence of Washington D.C: Citizens Plan for Prosperity in the 21st Century*, [21] which laid out a strategy to grow businesses, jobs, population, neighborhoods, and prosperity in the city. The plan called for the construction of a new Metro station along the Red Line near New York and Florida Avenues.

---

[20] Public-private partnerships and special assessment districts are described in more detail in Appendix B, Sections D-1 and E-2, respectively.

[21] District of Columbia Department of Housing and Community Development. *The Economic Resurgence of Washington, DC: Citizens Plan for Prosperity in the 21st Century.* November 1998.

Following the publication of this plan, the city's Department of Housing and Community Development created the New York Avenue Task Force to bolster economic development in the area and raise funds for the new Metro station. The task force obtained $350,000 in funding from the city to produce a feasibility study that examined the possibility and economic benefits of a Metro station at the intersection of New York and Florida Avenues. By connecting NoMa to the Metrorail network, the proposed station would connect NoMa to the entire region. The feasibility study projected that investing in the station would create 5,000 new jobs and $1 billion in private investment and development. These findings, along with an extensive outreach effort, helped the task force convince private landowners to provide $25 million, or about one-third of the estimated cost of the station. Remaining costs, which turned out to be higher than anticipated, were covered by the city and the federal government. The task force funded its extensive outreach efforts through a $100,000 grant from the city and a $140,000 contribution from the private sector.

FUNDING SOURCES AND FINANCING MECHANISMS

The New York Avenue Station was built with funds from private landowners, the city, and the federal government. Each party initially agreed to pay one-third of the cost, or $25 million, based on an initial total project cost estimate of $75 million. The city also paid cost overruns of over $25 million (Exhibit 3).

Private landowners agreed to pay a special assessment over 30 years to raise the $25-million private-sector contribution. This special assessment would be an additional charge on top of regular property taxes for nonresidential parcels within 2,500 feet of the future station's entrances. The city financed the project by issuing bonds that would be repaid using the funds collected through the special assessment.

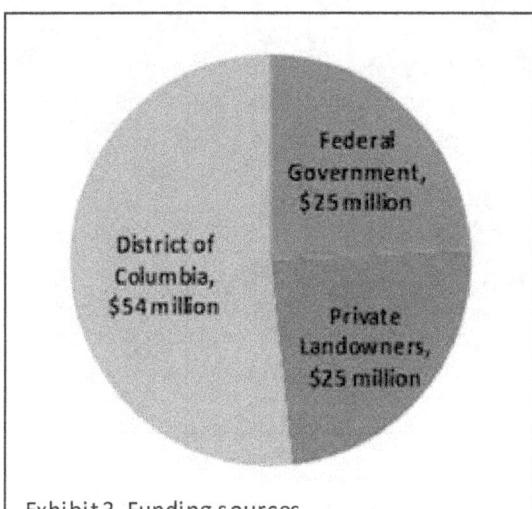

Exhibit 3. Funding sources.

Source: National Council for Public Private Partnerships. "New York Avenue Metro Station, Washington, DC." http://www.ncppp.org/cases/nystation.shtml. Accessed October 30, 2011.

The project components included the construction of the station and land acquisition around the station. The station did not require the construction of a new rail line to reach the area, reducing overall costs.

In March 2007, to continue making improvements in the area, the city formed the NoMa Business Improvement District (BID). An additional special assessment on commercial, multiunit residential, and hotel properties was levied to support community improvement investments, including cleaning and safety services, marketing, community development programs, and public events.

Private property owners were initially reluctant to participate, arguing that they would be paying twice, once through the assessment and once again through increased property taxes. After lengthy negotiations, property owners supported the BID's creation because of the anticipated increase in their property values. The BID's annual budget would be funded by the special assessment. The NoMa BID will expire in 2012, five years after its creation. However, the BID can be re-registered for additional five-year periods if the BID membership and the mayor approve after holding a hearing.

In 2008, the BID's annual budget was $1.3 million. The levy is structured as shown in Exhibit 4.

Property Type	Assessment
Land, parking lots, and industrial properties; properties less than 50,000 square feet	$0.05 per $100 of assessed value
Office and commercial properties over 50,000 square feet	$0.15 per rentable foot
Residential properties (10 or more units)	$120 per unit
Hotels	$90 per room

Exhibit 4. Levy Structure.

The federal government has also contributed to redevelopment in the area by locating federal offices in the business district, committing $100 million to build a headquarters office for the Bureau of Alcohol, Tobacco, and Firearms and another $100 million to build offices for the Securities and Exchange Commission. Such investments would not have occurred without the station.

LESSONS LEARNED

This project exceeded the predicted benefits in terms of jobs and investment. Assessed valuation of the 35-block area increased from $535 million in 2001 to $2.3 billion in 2011. It is estimated that over 15,000 jobs have been created with $1.1 billion in private investment (Exhibit 5). [22]

Exhibit 5. Newly completed and on-going construction in the NoMa neighborhood along the Metrorail line. Source: EPA.

KEYS TO SUCCESS

- The special assessment district allowed the project to proceed earlier than would have been possible using a pay-as-you-go approach because the District could issue bonds backed by projected future property tax revenue.

- Support from the city was essential to the project's success. Extensive stakeholder engagement ensured commitment to the project, even from those who were initially skeptical. The task force and the city succeeded in bringing the private sector to the table and ensuring that project costs were shared with those that would be benefiting from the increased property values.

KEY BARRIERS

- Significant time and resources were required to convince landowners of the potential benefits of a transit station and establish the special assessment district. Feasibility studies, an extensive public

---

[22] National Council for Public Private Partnerships. "New York Avenue Metro Station, Washington, DC." http://www.ncppp.org/cases/nystation.shtml. Accessed October 30, 2011.

outreach process, and coordination with multiple parties were critical components of implementation.

APPLICABILITY TO OTHER PLACES

This model could be implemented where the following conditions are met:

- A strong real estate market that attracts private investment.

- State law that allows the formation of special assessment districts.

Given that the transaction costs of implementing a special assessment district tend to be high, this tool is most appropriate for areas that need to raise significant amounts of capital because they have a single large project or several smaller projects that can be grouped together.

REFERENCES

District of Columbia Department of Housing and Community Development. *The Economic Resurgence of Washington, DC: Citizens Plan for Prosperity in the 21st Century.* November 1998. http://www.dcwatch.com/govern/dhcd9811.htm.

National Council for Public Private Partnerships. "New York Avenue Metro Station, Washington, DC." http://www.ncppp.org/cases/nystation.shtml. Accessed October 30, 2011.

FHWA Office of Innovative Program Delivery and AASHTO Center for Excellence in Project Finance. "Special Assessment Districts." http://www.transportation-finance.org/funding_financing/funding/local_funding/value_capture/special_assessment_districts.aspx. Accessed October 28, 2011.

FHWA Office of Innovative Program Delivery and AASHTO Center for Excellence in Project Finance. "New York Avenue-Florida Avenue-Gallaudet University Metro Center: A Case Study." Undated. http://www.transportation-finance.org/pdf/funding_financing/funding/local_funding/New_York_Avenue_Case_Study.pdf.

Washington Metropolitan Area Transit Authority. *New York Avenue-Florida Avenue Gallaudet University Station Access Improvement Study.* June 2010. http://www.wmata.com/pdfs/planning/NY%20Ave-FL%20Ave-Gall%20U%20Station%20Access%20Improvement%20Study%20Final%20Report.pdf.

The Maryland Transit Funding Study Steering Committee. *Committee Report.* January 2007. http://www.mdot.maryland.gov/Office%20of%20Planning%20and%20Capital%20Programming/Transit_Funding_Study/Documents/January2007CommitteeReport.pdf.

B. JOINT DEVELOPMENT: WEST DUBLIN BART STATION[23]

INTRODUCTION

The West Dublin Bay Area Rapid Transit (BART) Station opened in February 2011 after a nearly 15-year effort to fund and construct the station. It was the first infill station to be constructed in the BART system.[24] The construction of the West Dublin Station provides an example of fixed-guideway rail station construction and providing TOD infrastructure through joint development[25] and a paid parking strategy. Joint development, a form of value capture, is generally defined as a real estate development project that involves coordination between multiple parties to develop sites near transit, usually on publicly owned land.[26] As described in Chapter II, Section Value Capture, value capture tools capture some portion of the increased value or savings resulting from the public provision of new infrastructure. In the case of the West Dublin BART station, the joint development project captured a portion of the increased land value conferred by the new station.

PROJECT BACKGROUND

The West Dublin Station is in suburban Alameda County on the border between the towns of Dublin and Pleasanton in the San Francisco Bay Area (Exhibit 6). The station was completed as an infill station along a 10-mile stretch of the BART line where there previously were no stops, between the existing Dublin/Pleasanton and Castro Valley stations. During the original phase of railway implementation, BART planned to construct one large station in Dublin/Pleasanton. However in 1988, due to traffic flow concerns, the city of Dublin sued BART to mandate that BART diffuse the concentration of traffic by building two stations—one in the east and one in the west.

While BART had sought to construct a station since the late 1980s, the California Infrastructure Financing Act of 1996 allowed the actual first steps of project development. The act gave regional and local government entities like BART the ability to plan and

Exhibit 6. Platform 1 to Dublin/Pleasanton at the West Dublin/Pleasanton BART station.
Source: Eric Fischer via Flickr.com.

---

[23] A version of this case study appears in *Fullerton Smart Growth 2030: FTC Specific Plan Funding & Financing Strategy & Case Studies* prepared for City of Fullerton and Southern California Association of Governments by Strategic Economics. 2012.

[24] Lam, X. "New West Dublin/Pleasanton Station, BART's 44th, to open Feb. 19." *BART News Articles.*

[25] Joint development is described in more detail in Appendix B, Section E-4.

[26] Center for Transit-Oriented Development. *Capturing the Value of Transit.* Prepared for DOT and FTA. November 2008. p. 26.

enact development projects related to infrastructure improvement that would generate income or revenue for the agency. The act allowed much of the internal pre-project planning, such as identifying a station location, to occur before the environmental impact report was completed.[27] This legislation provided the framework for local agencies to create value capture funding mechanisms in concert with private-sector developers.

Two important factors influenced BART's choice for the station site. In anticipation of future infill station construction, BART had built an electrical substation near the West Dublin BART site during its initial development of the Dublin/Pleasanton extension. This project meant that a huge infrastructure component of the station was already in place. Second, and more importantly for the purposes of this report, BART owned several parcels in the selected station area that could be used to create value capture mechanisms to pay for the cost of the station.[28]

FUNDING SOURCES AND FINANCING MECHANISMS

VALUE CAPTURE IMPLEMENTATION

Besides being located in two municipal jurisdictions, the station area presented some unique development challenges. The station needed to be built in the median of a major freeway; as a result, two pedestrian bridges needed to be constructed for rider access to the station. BART also needed to provide parking for the park-and-ride patrons of a typical suburban station. The agency hoped to roll much of the cost of constructing these infrastructure improvements (including station construction) into a value capture strategy whereby BART would sell or ground-lease BART-owned parcels adjacent to the station to private developers.

BART's property acquisition team coordinated with the other government agencies whose jurisdiction overlapped the station area: the city of Dublin, the city of Pleasanton, and Alameda County's Surplus Property Authority. At the same time, BART solicited interest from private property developers for development in the station area on BART-owned land. BART ground-leased a 3-acre parcel to a group of private developers for 99 years for a one-time payment of $15 million.[29] BART and the developers also agreed to a covenant for a transit district transactional fee whereby a percentage of every sale of residential units in the development will be remitted to BART, allowing the agency to collect more revenue based on the level of development (i.e., the number of residential units and sale price of the land).

The development plan called for a transit village consisting of over 300 residential units, a hotel, and space for retail. However, in the wake of the housing and financial market crisis in 2008, development of the Dublin site stopped. The private developer was unable to meet project costs, and the project went into foreclosure. Although the site has not been developed, the right to develop was resold to a

---

[27] Personal communication with John Rennels, BART, by Eli Popuch, Strategic Economics, on October 10, 2011.

[28] Ibid.

[29] Ibid.

different development company at a sheriff's sale.[30] It remains a potential location for mixed-use TOD when the market for real estate returns.

In the nearby city of Pleasanton, another BART-owned parcel was originally zoned for commercial and office uses, but the BART property team was able to secure a change to residential and retail uses under a specific plan that the city was completing for the area. BART struck a similar ground-lease agreement with a private developer, this time with the developer paying $5 million in upfront costs. The developer plans to construct 350 residential units over 10,000 square feet of first-floor retail. Similar to the Dublin site, BART attached a covenant for a transit district transactional fee on the Pleasanton site. This provides BART with a guaranteed source of ongoing revenue from its properties, even after disposition, once development occurs. However, even though the BART station is open and operating, as of early 2012, no buildings had been developed on the leased parcels. Nevertheless, BART will be positioned to collect payments under the lease agreements in TOD projects that are completed.

STATION CONSTRUCTION

Once the private developer agreements were in place and a source of project funding secured, BART was able to begin constructing the station and adjacent infrastructure improvements. BART's property team secured approval for a general obligation bond from the BART Board of Governors. BART was willing to roll station construction costs into a larger systemwide bond in part because the parking garages built as part of the project implemented a paid parking strategy.

The parking fee revenue from the garages helped to make the project feasible and will help pay for operations. The garage on the Dublin side has 722 parking spaces, while the garage in Pleasanton holds 488 parking spaces.[31] The overall cost of the project was originally estimated at $87 million dollars but eventually rose to $106 million, due to complications with the pedestrian bridges. In the end, BART was able to apply the $20 million it had made in the land agreements to infrastructure improvements around the station.

LESSONS LEARNED

The West Dublin BART Station illustrates how to provide TOD infrastructure through joint development and a paid parking strategy. It also shows the importance of transit agencies working in concert with local governments and the private sector to ensure project completion. A significant amount of planning, financial leverage, and strategic investment can be required to implement a joint development project. Individually, BART, the cities of Dublin and Pleasanton, and the private development community did not have the resources to build a station with extensive infrastructure improvements and implement TOD. But when the interested parties came together and maintained frequent, open communication, the result was a new station that is primed for future development.[32] This point is especially true under weaker real estate market conditions, when the private sector is less inclined to invest in TOD projects that have higher construction costs and implementation challenges.

---

[30] Ibid.

[31] BART. "West Dublin/Pleasanton Station." http://bart.gov/stations/wdub/index.aspx. Accessed January 25, 2011.

[32] Personal communication with John Rennels op cit.

KEYS TO SUCCESS

- The value capture strategy enabled BART to use new real estate investment to finance infrastructure near the area where the development will occur. However, such a strategy can be feasible only in solid real estate markets where additional property tax or other revenue from new development is likely.

- Extensive multiagency coordination and unified support for a project can facilitate flexibility in planning and development requirements, making private investment possible in a challenging real estate market.

KEY BARRIERS

- One of the main challenges to using joint development as a value capture strategy is that it usually relies on revenue from a relatively small proportion of the property that is benefited by transit service. For instance, the private development portion of the West Dublin/Pleasanton project includes only about 20 acres, or about 4 percent of the property within a half-mile radius of the station. If a greater proportion of the property were involved in the development, or if a value capture strategy could draw from value increases from existing properties in the larger surrounding area, the potential for value capture would be greater.

- Large projects need multiple funding sources to succeed. No single source of funding can meet all TOD infrastructure needs.

APPLICABILITY TO OTHER PLACES

A value capture strategy using joint development would be most applicable where the following conditions are met:

- A strong real estate market.

- Strong multiagency and stakeholder coordination and support of a plan.

- A transit agency with sufficient real estate market knowledge and experience to enter into complex transactions.

REFERENCES

BART. "Home page." http://www.bart.gov. Accessed January 27, 2011.

BART. "West Dublin/Pleasanton Station." http://bart.gov/stations/wdub/index.aspx. Accessed January 25, 2011.

Center for Transit-Oriented Development. *Capturing the Value of Transit*. Prepared for DOT and FTA. November 2008. http://www.reconnectingamerica.org/assets/Uploads/ctodvalcapture110508v2.pdf.

Lam, Xuan "New West Dublin/Pleasanton Station, BART's 44th, to open Feb. 19." *BART News Articles*. http://www.bart.gov/news/articles/2011/news20110121.aspx.

Personal communication with John Rennels, BART, by Eli Popuch, Strategic Economics, on October 10, 2011.

C. FEDERAL LOANS, GRANTS, AND CREDIT ENHANCEMENT: DENVER UNION STATION

INTRODUCTION

Denver Union Station is a $500-million, multiparty, multijurisdictional redevelopment project supported by local and federal funding and financing sources. This project marks the first time that federal loans from the Railroad Rehabilitation and Improvement Financing (RRIF) and Transportation Infrastructure Finance and Innovative Act (TIFIA) programs have been combined to fund a major infrastructure project.[33]

PROJECT BACKGROUND

Denver Union Station is in lower downtown Denver (Exhibit 7). During its heyday in the 1940s, the station was served by 80 daily trains operated by six different railroads. Today, however, only one train a day stops at the historic station. In 2001, the Regional Transportation District (RTD), the city and county of Denver, the Colorado Department of Transportation (CDOT), and the Denver Regional Council of Governments (DRCOG) signed an intergovernmental agreement to consider various redevelopment options for the station. In accordance with the agreement, RTD purchased the station, and RTD, the city and county, CDOT, and DRCOG (the "project partners") jointly initiated the Denver Union Station Master Plan.[34] After an extensive public outreach process and environmental review, the project partners approved the final master plan in 2004.

In November 2006, the project partners selected Union Station Neighborhood Company as the master developer to head the redevelopment project. In July 2008, the Denver City Council created the Denver Union Station Project Authority (DUSPA), which is responsible for financing, acquiring, owning, equipping, designing, constructing, renovating, operating, and maintaining the Denver Union Station redevelopment project.[35] DUSPA serves as the financing entity for the Denver Union Station project and the contracting entity for the construction contracts.

As conceived by the project partners and the master developer, the project will transform the site—including the

Exhibit 7. Denver Union Station.
Source: K_Gradinger via Flickr.com.

---

[33] The RRIF and TIFIA programs are described in more detail in Appendix B, Sections B-9 and C-2, respectively.

[34] DUSPA. "Denver Union Station History and Timeline." http://www.denverunionstation.org/index.php?option=com_content&view=article&id=3&Itemid=4. Accessed August 22, 2012.

[35] Mancini Nichols, C. "Value Capture Case Studies: Denver's Historic Union Station." Metropolitan Planning Council. April 19, 2012. http://www.metroplanning.org/news-events/article/6392.

historic Union Station building, rail lines, adjacent vacant parcels, street rights-of-way, and offsite trackage rights (adjacent to rail tracks)—into an intermodal transportation district surrounded by TOD, including 280,000 square feet of residential space, 70,000 square feet of retail property, and 1 million square feet of office and hotel development. The transit district will connect commuter rail, light rail, bus rapid transit, [36] regular bus service, and other transportation services across the region. [37] Project components include:

- Construction of light rail and commuter rail stations.
- A 22-bay underground regional bus station.
- Extension of the 16th Street Mall and the associated shuttle service.
- Accommodation of the Downtown Circulator bus service.
- Street improvements.
- Parking replacement. [38]

Nine years after the initial planning phase, the project began construction in February 2010. It is expected to be completed in 2014.

The project's transit elements are expected to cost approximately $500 million (estimates ranged from $434.5 million in the final environmental impact statement to $518.6 million in the TIFIA application). Exhibit 8 shows the breakdown of the project costs, as presented in the final environmental impact statement.

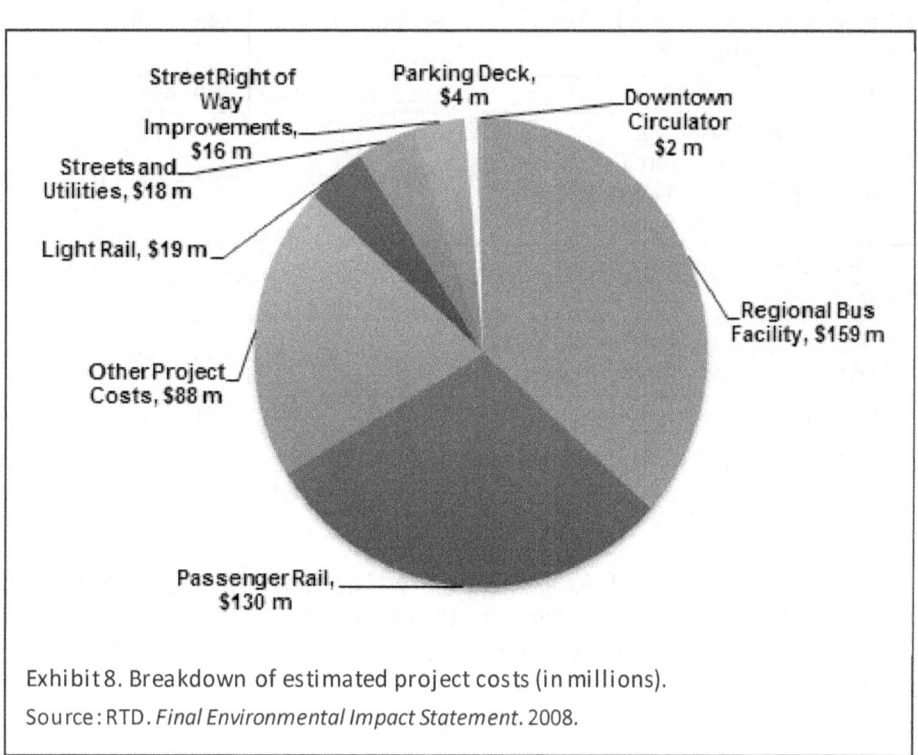

Exhibit 8. Breakdown of estimated project costs (in millions).

Source: RTD. *Final Environmental Impact Statement*. 2008.

---

[36] Bus rapid transit is a public transportation system with improved infrastructure, vehicles, and scheduling designed to enable buses to provide faster and more efficient service.

[37] FHWA Office of Innovative Program Delivery. "Project Profile—Denver Union Station." 2010.

[38] Ibid.

FUNDING SOURCES AND FINANCING MECHANISMS

The project partners used a combination of federal and local sources to finance the project. Federal loans from TIFIA and RRIF provided $300 million in loans at a low interest rate[39] to finance the bus and rail facilities, allowing the project partners to provide the transportation choices they envisioned and facilitating plans for private development on the land surrounding the site. This project marks the first time TIFIA and RRIF have been used together to fund a major infrastructure project.

The remaining $200 million in projects costs will come from revenue and grants. Exhibit 9 summarizes the funding and financing sources.

Source	Type	Amount	Term/Note
RRIF	Federal loan	$155 million	30-year, under 4%. For rail infrastructure
TIFIA	Federal loan	$145.6 million	30-year, under 4%. For the bus station
Revenue during construction	Project revenue	$57.5 million	
Federal Highway Administration Projects of National and Regional Significance grant	Federal grant	$45.3 million	
RTD contribution	Local funding	$40 million	
American Recovery and Reinvestment Act grant (Urbanized Area Formula and Federal Highway Administration Flex Funds)	Federal grant	$28.4 million	For construction of station project and rolling stock
Other state and local funds	Local funding	$19.9 million	
Land sales	Project revenue	$17.4 million	
Federal Transit Administration Transit Capital Investment Program grant	Federal grant	$9.5 million	
Total funding		**$518.6 million**	

Exhibit 9. Funding and financing sources.
Source: Arup based on information from FHWA Office of Innovative Program Delivery. "Project Profile—Denver Union Station." 2010.

To repay the $145 million TIFIA loan, RTD agreed to pay DUSPA $12 million annually for 30 years. Two sources will repay most of the project debt: an RTD bond secured by gross sales tax revenue, which will repay the TIFIA loan, and tax-increment financing, which will repay the RRIF loan.

To be eligible for TIFIA loans, a project must receive an investment grade rating on its project debt from a Wall Street rating firm. Several factors helped the project secure a Fitch rating of "A," an investment grade, on the project's TIFIA debt obligation, including:

[39] Slightly under 4 percent, well below the rate that DUSPA could have secured in the tax-exempt capital market.

- RTD's pledge to pay $12 million annually is a multiyear obligation that is not subject to annual action by RTD or DUSPA, which eliminates any risk that the agencies will fail to appropriate money to pay back the TIFIA debt.

- RTD payments have a designated funding source, a 0.4 percent FasTracks sales and use tax approved by voters in 2004.[40]

- In its 2009 financial projection, RTD shows the debt service coverage ratio[41] for all existing and proposed debt falls between 3.76 percent (in 2016) and over 9 percent (after 2019). This value implies that if the project revenue comes in slightly lower than anticipated, RTD will likely still be able to meet its TIFIA debt obligation.

Revenue from a TIF[42] district that encompasses the 40-acre commercial development around the station is the primary source that will be used to repay the RRIF debt. The Denver Downtown Development Authority has pledged to use TIF revenue for 30 years to secure and repay the RRIF debt obligation. Because TIF revenue is speculative—it is based on anticipated increased property value caused by the project—the RRIF loan is backed by a "moral obligation" from Denver, meaning that in the event of a shortfall in revenue for debt service on the RRIF loan, the city will appropriate up to $8 million annually from the general fund to make up the shortfall. Revenue from a special assessment and a local hotel tax will also contribute to paying back the RRIF loan.

LESSONS LEARNED

The Denver Union Station model is suitable for large TOD projects because TIFIA, one of the primary financing tools used, is available only for projects with costs exceeding $50 million and a dedicated non-federal revenue source. RRIF funds are limited to the rail itself or related facilities and therefore could not be used for TOD infrastructure as defined in this report. However, RRIF is an example of a funding and financing tool that could contribute to making TOD infrastructure projects possible by funding the transit and thereby freeing up other funds that could be applied to the TOD infrastructure.

KEYS TO SUCCESS

- Access to capital markets helped expedite project delivery. DUSPA was able to rely on future property and sales tax revenue to support two federal loans (TIFIA and RRIF) that allowed the project partners to implement the project earlier than would have been possible using a pay-as-you-go approach.

- The project partners were able to use value capture mechanisms (e.g., TIF, special assessment district) to help pay for the project. Value capture mechanisms are most feasible in solid real estate markets where new development is likely to occur, generating additional property tax revenue. Because it was difficult to predict how development would proceed in downtown Denver given the

---

[40] FHWA Office of Innovative Program Delivery. "Project Profile—Denver Union Station." 2010.

[41] Debt service coverage ratio is the ratio of cash available for debt servicing to interest, principal, and lease payments. It is a popular benchmark used to measure an entity's ability to produce enough cash to cover its debt (including lease) payments. The higher this ratio is, the easier it is to obtain a loan.

[42] More information on TIF is provided in Appendix B, Section E-3.

economic conditions, the city and county of Denver agreed to take a moral obligation to obtain the RIFF loan, agreeing to appropriate additional funds if TIF revenue does not fully cover debt service payments.

- Extensive multiagency coordination and unified support for a project can facilitate access to capital markets and expedite project delivery. The Denver Union Station project has a regional impact, benefiting not only the city of Denver but also the larger metropolitan area. It was only through the coordination and support of multiple agencies that DUSPA was able to secure both TIFIA and RRIF loans.

KEY BARRIERS

- Federal loans such as TIFIA and RRIF have become increasingly competitive in recent years. Securing these loans requires demonstrating the security of the repayment scheme. Project sponsors should consult with a financial advisor early during the development process to design repayment schemes that will be acceptable to federal loan issuers. Both TIFIA and RRIF federal loans require security for their respective debt repayment schemes from the loan applicant. For the TIFIA loan, DUSPA needed to give TIFIA the senior lien for the project revenue. In other words, TIFIA debt obligations will be fulfilled before other loans receive repayments from the project revenue. In addition, DUSPA needed to structure the repayment scheme so that it would not be subject to future appropriation approvals, effectively eliminating a major repayment risk. For the RRIF loan, DUSPA needed to obtain a pledge from the city to secure and repay the RRIF loan under a moral obligation.

- Project planning and implementation were time intensive; working with multiple agencies requires time, coordination, and negotiations, and developing a TOD project is also a lengthy process. The Denver Union Station redevelopment concept originated in 2001 with RTD's purchase of the station. After completing master planning, environmental approvals, and public hearings; selecting a developer; and securing funding sources, the project finally began construction in 2010, nine years after the first phase of the redevelopment effort. Starting the development process early will increase the likelihood of completing the project on time.

APPLICABILITY TO OTHER PLACES

This model could be implemented where the following conditions are met:
- Strong multiagency coordination and unified support of a project.
- Large-scale projects.
- A strong real estate market that can attract private investment.

REFERENCES

Barrett, Diane S. "Financing Denver Union Station." City of Denver. June 2011. http://www.iscvt.org/where_we_work/usa/article/low_carbon_transportation/barrett_denver.pdf.

DUSPA. "Board Meeting Materials." November 12, 2009. http://www.denverunionstation.org/index.php?option=com_content&view=article&id=59&Itemid=65.

DUSPA. "Denver Union Station History and
Timeline." http://www.denverunionstation.org/index.php?option=com_content&view=article&id=3&Itemid=4.
Accessed August 22, 2012.

DUSPA. "DUSPA Meeting
Documents." http://www.denverunionstation.org/index.php?option=com_content&view=article&id=59&Itemid=6
5. Accessed August 22, 2012.

DUSPA. "Denver Union Station Project Authority Closes Two Loans with U.S. Department of Transportation." July
23, 2010. http://www.rtd-fastracks.com/media/uploads/main/DUSPA_loan_closing_72310.pdf.

DUSPA. "Fact Sheets." http://www.denverunionstation.org/index.php?option=com_content&view=article&id=52.
Accessed October 27, 2011.

DUSPA. "Fitch gives DUS investment grade designation." November 12,
2009. http://www.denverunionstation.org/index.php?option=com_content&view=article&id=80&Itemid=80.

DUSPA. "Press
Releases." http://www.denverunionstation.org/index.php?option=com_content&view=article&id=80&Itemid=80.
Accessed August 22, 2012.

DUSPA. "Summary of Documents for Distribution for DUSPA Finance Committee Meeting." January 19,
2010. http://www.denverunionstation.org/index.php?option=com_content&view=article&id=59&Itemid=65.

FHWA Office of Innovative Program Delivery. "Project Profile—Denver Union Station."
2010. http://www.fhwa.dot.gov/ipd/project_profiles/co_union_station.htm.

Jackson, Margaret. "Laying the rails for Union Station." *The Denver Post.* January 31, 2010.

Leib, Jeffrey. "Union Station construction kept off track. The redevelopment project may not start on time because
a rating holdup is stopping the dominoes from falling." *The Denver Post.* October 27, 2009.

Mancini Nichols, Chrissy. "Value Capture Case Studies: Denver's Historic Union Station." Metropolitan Planning
Council. April 19, 2012. http://www.metroplanning.org/news-events/article/6392.

Nelson, Ketrina. "FTA's Peter Rogoff announces $304 million for Denver Union Station." Federal Transit
Administration. February 5, 2010.

RTD. *Final Environmental Impact Statement (Chapter 7—Financial Assessment).*
2008. http://www.denverunionstation.org/index.php?option=com_content&view=article&id=47&Itemid=50.

D. SPECIAL ASSESSMENT DISTRICT: DOWNTOWN STAMFORD

INTRODUCTION

Stamford, Connecticut's downtown revitalization effort is an example of multiagency coordination with strong private-sector support. The creation of the Stamford Downtown Special Services District (DSSD) in 1992 helped catalyze the effort. Stamford's example illustrates how resources from special assessments[43] and private sponsorships can be used over the long term to achieve TOD goals, including promoting an active and pedestrian-friendly environment and encouraging a vital downtown district served by commuter rail.

PROJECT BACKGROUND

Stamford is in Fairfield County, Connecticut, on the main branch of the New Haven Line of the Metro-North Railroad, the commuter rail system for northern metropolitan New York City. Stamford has a diverse economic base and is the business center of Fairfield County. Many major U.S. companies have corporate headquarters in Stamford.

From the late 1980s through the early 1990s, downtown Stamford struggled with high office and retail vacancy rates, a weak economy, dilapidated infrastructure, and few amenities and services for residents and visitors. In 1992, downtown property owners voted overwhelmingly to create a business improvement district, called the downtown special service district, to manage and revive the downtown. Various agencies have been instrumental in the revitalization, including:

- DSSD, which works with the city to promote the downtown.

- The city's Office of Economic Development, which works to attract and retain businesses.

- The Planning Board and the Urban Redevelopment Commission, which oversee development and redevelopment of land and buildings.[44]

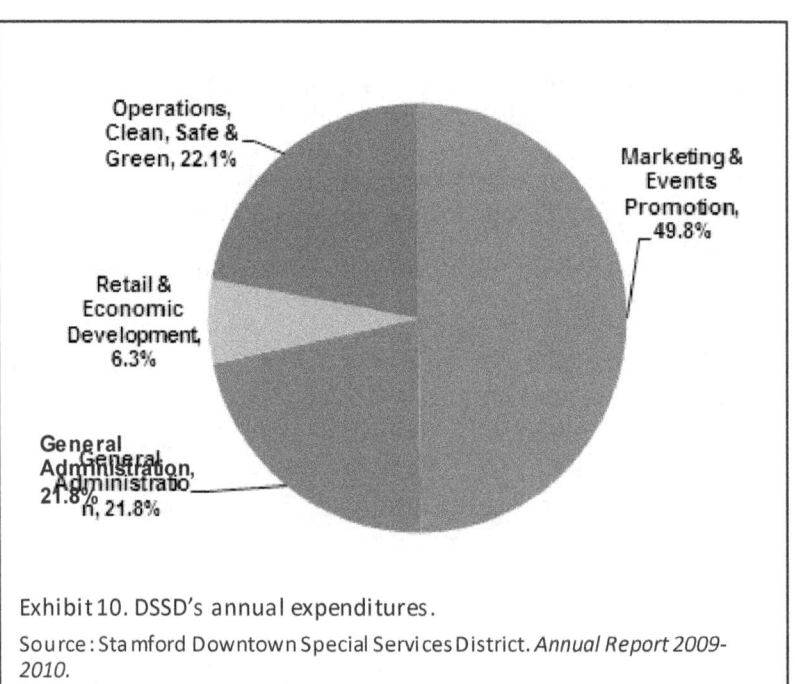

Exhibit 10. DSSD's annual expenditures.

Source: Stamford Downtown Special Services District. *Annual Report 2009-2010.*

---

[43] Special assessments are discussed in greater detail in Appendix B, Section E-2.

[44] City of Stamford. "Stamford's Business Climate." http://www.cityofstamford.org/content/25/50/258/2753/default.aspx. Accessed August 20, 2012.

DSSD focuses on three main areas (Exhibit 10):

- Attracting people to the downtown. Marketing and events promotion are major activities, receiving nearly 50 percent of DSSD's resources. Events include outdoor cultural exhibits, parades, concerts, and a farmers' market.

- Enhancing the downtown's outdoor environment. DSSD operates a Clean Team to control litter; Downtown Ambassadors to enhance neighborhood safety; a Green Team to take care of the streets, landscape, and parks; Streetscape Operations to work on maintenance issues; and a sidewalk snow removal program.[45]

- Bolstering downtown's economic development. DSSD seeks to ensure new development meets environmental and quality of life objectives. Using Stamford's master plan as a guide, DSSD has helped foster a downtown revival with retail, office, cultural, recreation, and residential uses. DSSD is working with the Urban Redevelopment Commission and the community to produce a downtown master plan, which will be followed by corresponding rezoning.[46]

FUNDING SOURCES AND FINANCING MECHANISMS

DSSD is a type of special assessment district, also referred to as a business improvement district, authorized by Connecticut law. Under Connecticut statute, a municipality can form a DSSD to promote the economic and general welfare of its citizens and property owners. "Among other things, the district can:

- acquire and convey real and personal property;

- provide any service that a municipality can provide, other than education;

- recommend to the municipality's legislative body that it impose a separate tax on property in the district to support its operations;

- borrow money for up to one year backed by district revenue; and

- build, own, maintain, and operate public improvements."[47]

Stamford's DSSD is funded primarily by two sources: special tax assessments on downtown property owners and private sponsorships (see Exhibit 11). Approximately 128 property owners pay an annual fee to DSSD based on their property values. The district's annual resources are close to $4 million.

---

[45] Stamford Downtown Special Services District. *Annual Report 2010-2011.*

[46] Ibid.

[47] Pinho, R. "Special Services Districts." *OLR Research Report.* June 25, 2012.

LESSONS LEARNED

Stamford's model shows how revenue from special assessments on downtown property owners can be used over the long term to operate and maintain pedestrian infrastructure and encourage vitality in a downtown.

KEYS TO SUCCESS

- DSSD actively pursues comprehensive strategic planning for the downtown over the long term.

- Small actions such as beautifying streets and marketing and events promotion helped create momentum for the revitalization efforts.

- DSSD's special assessment provides a long-term funding source for operations and maintenance of downtown infrastructure. Funding for operations and maintenance of capital improvement projects is often overlooked.

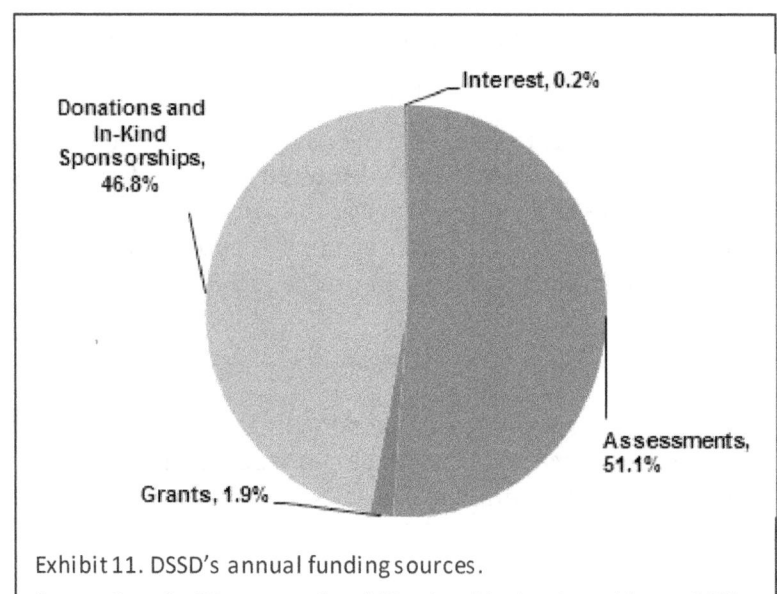

Exhibit 11. DSSD's annual funding sources.

Source: Stamford Downtown Special Services District. *Annual Report 2009-2010.*

- The Stamford model relies strongly on private support through special assessments and private sponsorships. This model is possible because of the large number of corporations located in the city, and the fact that property values have increased since the inception of the DSSD.

KEY BARRIERS

- Special assessment districts are typically subject to voter approval. Outreach to and education of property owners are therefore critical in implementing special assessment districts.

APPLICABILITY TO OTHER PLACES

This model could be implemented where the following conditions are met:

- Strong multiagency coordination and unified support of a project.

- State law allows for special assessments and enables special districts to provide services that a municipality would otherwise provide. In this case, Connecticut law allows DSSD to perform duties that are generally a city or county's responsibility, such as street cleaning and patrolling.

Though the Stamford model focuses mainly on maintaining pedestrian infrastructure and promoting the downtown area, the resources collected through property taxation or sponsorships could also be used for small capital investments (less than $4 million, given DSSD's annual revenue), since Connecticut law provides districts enough flexibility to use their resources for either capital investments or maintenance expenses. Other areas interested in replicating the Stamford model would need to give their districts enough flexibility to adjust their priorities (maintenance vs. capital investments) based on their needs.

REFERENCES

Stamford Downtown Special Services District. *Annual Report 2006-2007.* http://www.stamford-downtown.com/UserFiles/File/pdf_AR2006-2007.pdf

Stamford Downtown Special Services District. *Annual Report 2009-2010.* http://www.stamford-downtown.com/UserFiles/File/AnnualReports/pdf__2009-10DSSDAR.pdf

Stamford Downtown Special Services District. *Annual Report 2010-2011.* http://www.stamford-downtown.com/UserFiles/File/AnnualReports/pdf_2010-11DSSDannualreport_web.pdf.

City of Stamford. "Economic Development Annual Reports." http://www.cityofstamford.org/content/25/50/258/3189/default.aspx. Accessed November 21, 2011.

City of Stamford. "Stamford's Business Climate." http://www.cityofstamford.org/content/25/50/258/2753/default.aspx. Accessed August 20, 2012.

Pinho, Rute. "Special Services Districts." OLR Research Report. June 25, 2012. http://www.cga.ct.gov/2012/rpt/2012-R-0287.htm.

E. PUBLIC-PRIVATE PARTNERSHIP: THE NEW QUINCY CENTER

INTRODUCTION

The New Quincy Center in Quincy, Massachusetts, is an example of a public-private partnership in which the private sector or developer bears the construction, design, and financial risks of developing TOD infrastructure. The city then reimburses the developer through taxes captured by a special assessment district on new development. However, the city will proceed with reimbursements only when an occupancy threshold has been achieved to ensure that income from property taxes from new development will be enough to reimburse the developer. Public-private partnerships for TOD can take many forms depending on the real estate market conditions and the project's scale.[48]

PROJECT BACKGROUND

The New Quincy Center is a transit-oriented, master-planned, mixed-use development that includes the redevelopment of approximately 50 acres of downtown Quincy. The city of Quincy (population about 92,000) is eight miles south of Boston and offers access to major freeways and to the region's public transportation system.[49]

Exhibit 12. People walking from the Quincy transit center towards downtown.
Source: EPA.

The city of Quincy spent three years negotiating an agreement with a company to redevelop the downtown. In January 2011, the city and a joint venture controlled by the developer[50] executed a land disposition agreement that created an innovative public-private partnership to redevelop downtown Quincy. The construction of the project is expected to start in the second quarter of 2012.

Over 2.7 million square feet of private, mixed-use buildings, including retail, office, and residential space, will be added to the center of Quincy (see Exhibit 13). The proposed schedule for the redevelopment has four phases, and its completion is anticipated eight years after it starts. The mix of uses and

---

[48] Public-private partnerships and special assessment districts are described in more detail in Appendix B, Sections D-1 and E-2, respectively.

[49] Newmark Grubb Knight Frank. "Private/Public Partnership Thrives in Quincy, MA, $1.6B Urban Revitalization Project." June 27, 2011.

[50] Minority partners of the joint venture include a local insurance firm, an Atlanta real estate group, a retail real estate owner/developer in the Northeast, and others.

timing of the redevelopment can be adjusted to respond to market opportunities as they arise.

	Office	Retail	Hotel		Residential		Total	Parking
	Sq. Ft.	Sq. Ft.	Sq. Ft.	Rooms	Sq. Ft.	Units	Sq. Ft.	Spaces
Step 1	86,850	260,590	68,120	136	101,540	74	517,100	718
Step 2	441,375	151,925	72,660	145	366,190	267	1,032,150	1283
Step 3	486,160	143,370	-	0	220,805	161	850,335	1332
Step 4	-	15,395	-	0	320,110	233	335,505	235
	1,014,385	571,280	140,780	281	1,008,645	735	2,735,090	3,568

Exhibit 13. Development program by land use for New Quincy Center.
Source: Street-Works Development LLC. "Exhibit 'B' to the LDA: The Development Plan." 2010.

The public improvements that will serve the redevelopment include utilities renovation, roads, sidewalks, street trees, landscaping, and public parking. The initial cost estimate was $277 million, of which $50 million will be financed by the city through state and federal funds, with the rest financed by the developer through equity and debt.[51]

The land disposition agreement structures the planning, execution, and financing of the public improvements associated with the redevelopment project. According to the agreement:

- The developer is responsible for designing, permitting, and constructing public improvements that specifically serve the redevelopment using private financing.

- The city of Quincy will purchase the public improvements related to each phase once certain conditions are met. Conditions related to the city's purchase of non-parking public infrastructure (e.g., utilities, sidewalks, and landscape) include:

  o Completion of non-parking public improvements.

  o Property tax payments for at least two quarters.

  o A bond or escrow account equal to 150 percent of the estimated cost to complete each construction task.

Conditions related to the city's purchase of parking include:

- Substantial completion of public parking improvements.

- Property tax payments for at least two quarters.

- 75 percent occupancy of residential, office, and other buildings.[52]

This model allocates the financing risk during the construction of the public improvements to the private partner, its lenders, and investment partners. In addition, the private partner assumes "occupancy risk"

---

[51] Street-Works Development LLC. "Exhibit 'B' to the LDA: The Development Plan." 2010.

[52] W-ZHA, LLC. "Summary of Financial Presentation to the Quincy City Council." 2010.

for the development's private uses, meaning that if the real estate occupancy threshold is not achieved, the city will not reimburse the developer for the public improvements.

By establishing a minimum occupancy threshold that needs to be met before the city purchases the public infrastructure, the city ensures that the new development will generate enough property tax revenue to pay the debt service of the tax-exempt general obligation bonds issued to finance the purchase.

FUNDING SOURCES AND FINANCING MECHANISMS

The city's public infrastructure will be upgraded in the short term by using private money and state and federal funds. Then, under the conditions summarized above, the city of Quincy will purchase the public improvements with the proceeds of tax-exempt general obligation bonds. Two revenue streams will be considered when assessing the bonding capacity:

- Net parking garage revenue.

- An increase in property taxes through special assessments as regulated by Massachusetts General Laws Chapter 121A. It is estimated that residential units will pay $4.50 per square foot while commercial properties' contributions will vary between $9.50 and $10.50 per square foot.[53]

Massachusetts General Laws Chapter 121A (121A) sets forth procedures for negotiating an alternative tax payment on certain developments. The 121A agreement appended to the land disposition agreement for the New Quincy Center sets an alternative tax assessment that authorizes higher rates than current ad valorem real estate taxes. The 121A payments will provide the city with a revenue stream from the redeveloped properties in each phase. These payments will be made to the city each quarter starting at substantial completion of each building.[54] During project construction before the 121A payments begin, the developer will make payments to the city to cover what it would normally receive in real estate taxes, net revenue from public parking facilities, and other property taxes.[55]

The maintenance and repairs of the new public infrastructure over a 30-year period will be supported by the District Improvement Financing Maintenance Fund, to which new development will contribute.[56] The base payment for each phase of development will be 50 cents per gross square foot of building area, adjusted for inflation from the date of the land disposition agreement. The payments will escalate at 12.5 percent every five years. These payments will be collected each quarter for 30 years, regardless of any termination of the 121A agreement. The operating expenses for parking will be funded by parking garage revenue.

---

[53] City of Quincy. *Land Disposition Agreement for the Quincy Center Redevelopment Project.* 2011.

[54] Ibid.

[55] Fishman, R. "Thinking outside the box to build a square: Innovative financing fuels Quincy Center redevelopment." *Commonwealth.* April 14, 2011.

[56] City of Quincy. *Land Disposition Agreement for the Quincy Center Redevelopment Project.* 2011.

LESSONS LEARNED

The New Quincy Center is a model for financing the construction of TOD infrastructure and maintaining it over the long term by establishing a public-private partnership.

KEYS TO SUCCESS

- The model provides infrastructure to support the redevelopment of downtown Quincy while insulating the public sector from the risk that the property tax revenue from the new development might not cover the full cost of the infrastructure. The private sector is responsible for the entire redevelopment, including design, permitting, financing, and execution, and thus bears the related risks. The city can get the infrastructure it needs more quickly and pay for the infrastructure if and only if the new development generates a certain level of tax revenue.

- The long-term maintenance of the public infrastructure built by the developer will be secured by the establishment of a maintenance fund to which property owners will contribute.

KEY BARRIERS

- Public-private partnerships tend to require long-term negotiations that have high transaction costs. Government authorities must have the patience and resources to engage in such a process for it to be successful.

- The model relies heavily on property taxes derived by new development. Thus, it requires large sites or projects and a solid real estate market to be viable.

- The Quincy model relies on alternative taxation on property through legislative authority that might not be available in other states.

APPLICABILITY TO OTHER PLACES

A public-private partnership in which the private sector or developer bears the construction, design, and financial risks of developing TOD infrastructure can be appropriate where the following conditions are met:

- Public infrastructure needs extensive upgrades.

- City or state regulation allows ownership of public infrastructure by private entities.

- The local real estate market is strong enough to attract significant private investment.

The mechanism used in the New Quincy Center example relies on the Massachusetts General Laws Chapter 121A. This law allows alternative taxation for real estate projects. For instance, Chapter 121A allows a municipality to offer a tax break that encourages private initiative to develop property in areas with high property tax rates or in areas that have difficulty attracting private investment. Analogous authority might not exist in other places.

REFERENCES

Adams, Steve. "Quincy Center Makeover One of Most Ambitious in New England." *The Patriot Ledger.* February 6, 2012. http://www.patriotledger.com/topstories/x2124806939/Quincy-Center-makeover-dwarfs-most-other-New-England-redevelopments.

City of Quincy. *Land Disposition Agreement for the Quincy Center Redevelopment Project.* 2011. http://www.quincyma.gov/Government/PLANNING/LandDispositionAgreement.cfm.

City of Quincy. *Quincy Center District Urban Revitalization and Development Plan.* 2007. http://www.quincyma.gov/CityOfQuincy_Content/documents/URDP%20Final-4-30-07%20v.5%20with%20cover(May-31-2007).pdf.

Fishman, Robert. "Thinking outside the box to build a square: Innovative financing fuels Quincy Center redevelopment." *Commonwealth.* April 14, 2011. http://www.commonwealthmagazine.org/Voices/Perspective/Online-Exclusives-2011/Spring/Thinking-outside-the-box-to-build-a-square.aspx.

Newmark Grubb Knight Frank. "Private/Public Partnership Thrives in Quincy, MA, $1.6B Urban Revitalization Project." June 27, 2011. http://www.newmarkkf.com/home/media-center/press-releases.aspx?d=4684.

Stephenson Design Group. "New Quincy Center Redevelopment." http://sdg-eng.com/projects/new_quincy_center_redevelopment. Accessed October 27, 2011.

Street-Works Development LLC. "Exhibit 'B' to the LDA: The Development Plan." 2010. http://www.quincyma.gov/CityOfQuincy_Content/documents/Tab%2002%20Exh%20B-Development%20Plan.pdf.

W-ZHA, LLC. "Summary of Financial Presentation to the Quincy City Council." 2010. http://www.quincyma.gov/CityOfQuincy_Content/documents/Tab%2009%20Financial%20Presentation%20to%20City%20Council.pdf.

F. SPECIAL TAX AND DENSITY INCENTIVES: THE WHITE FLINT SECTOR PLAN[57]

INTRODUCTION

The White Flint Sector Plan in Montgomery County, Maryland, is a comprehensive plan for districtwide TOD infrastructure improvements that is being implemented with a combination of value capture mechanisms. While creating the plan, the county explored the potential for using some combination of transportation impact fees, a special tax, and/or TIF to fund infrastructure.[58] However, only the special tax is likely to be implemented. The county is also using density incentives to encourage developers to provide some public facilities, including parking. The funding and financing tools discussed in this case study are most applicable to districts (as opposed to individual projects) with major infrastructure needs, where the real estate market is strong enough to support significant new development.

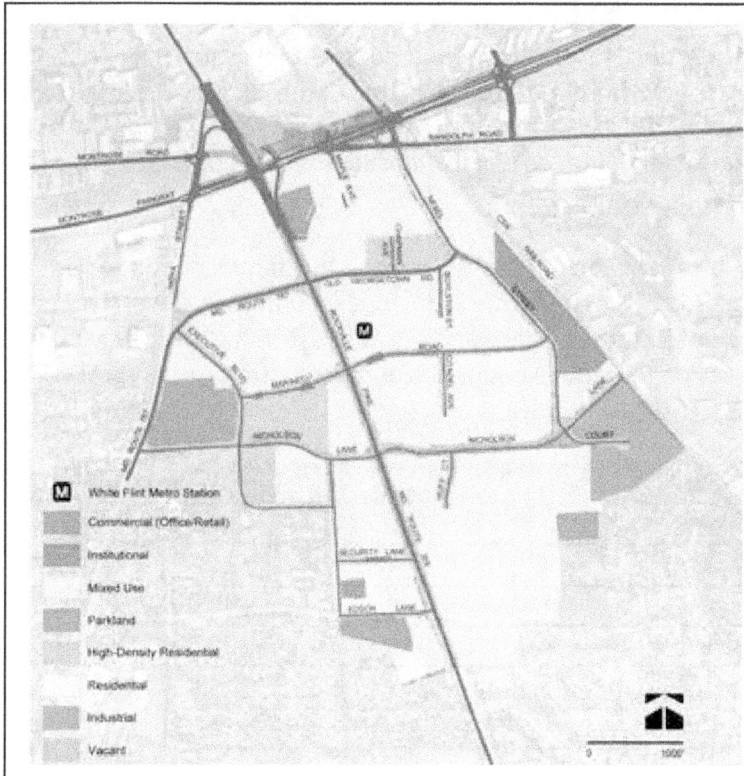

Exhibit 14. White Flint Sector Plan area.
Source: Montgomery County Planning Department, M-NCPPC. *White Flint Sector Plan.* 2010

PROJECT BACKGROUND

White Flint is in North Bethesda, an unincorporated but urbanized area in Montgomery County. The White Flint Sector Plan (Exhibit 14) covers 430 acres that lie within a three-quarter-mile radius of the White Flint Metrorail Station (Exhibit 15).

The White Flint district has large parcels under relatively consolidated ownership, including a regional shopping mall, parking lots, and strip malls that present opportunities for large-scale infill development. Since the White Flint Metrorail station was built in the 1970s, Montgomery County's long-range plans have identified the district as a place to accommodate a substantial portion of the region's projected growth, especially housing.[59] The most recent planning efforts were spurred in part by increasing congestion on

---

[57] A version of this case study appears in *Fullerton Smart Growth 2030: FTC Specific Plan Funding & Financing Strategy & Case Studies* prepared for City of Fullerton and Southern California Association of Governments by Strategic Economics. 2012.

[58] Impact fees, special taxes, and TIF are described in more detail in Appendix B, Sections E-1, E-2, and E-3, respectively.

[59] Montgomery County Planning Department, M-NCPPC. *White Flint Sector Plan.* 2010.

Exhibit 15. White Flint Metrorail Station parallels Rockville Pike (on the right).
Source: EPA.

Rockville Pike, a six-lane arterial that bisects the plan area and connects Montgomery County and other northern suburbs to Washington, D.C. The plan is intended to improve the pedestrian environment, while adding a significant amount of new housing to help balance land uses in the corridor, which is now largely commercial.

The sector plan process was led by the Maryland-National Capital Park and Planning Commission (M-NCPPC)[60] and involved a diverse group of community and private-sector stakeholders, including a coalition of major property owners and developers. The plan, approved by the Montgomery County Council in 2010, recommends the approval of 9,800 new housing units and 5.8 million square feet of new commercial space, on top of the 4,600 residential units and 7.3 million commercial square feet that were either in existence or approved for development as of 2010. The plan recommends siting the additional homes near transit facilities, including the existing Metrorail station and a planned commuter rail station.

The plan states that "implementing the White Flint Sector Plan will require substantial public and private investment in infrastructure and public facilities."[61] Major projects include reconfiguring local streets and intersections; building a new commuter rail station, police and fire and rescue facilities, a civic green, and a library; and transforming Rockville Pike into a boulevard with street trees, underground utilities, and dedicated lanes for bus rapid transit (Exhibit 16).

Infrastructure Project Type	Estimated Cost (Millions)
Local streets and intersections	$90.3
Rockville Pike Boulevard	$81.6
Streetscape improvements	$42.0
Market Street and promenade	$28.3
Metrorail northern station entrance	$25.0
Police and fire/rescue	$19.8
Commuter rail station/access improvements	$15.0
Civic green	$6.5
Library	$5.0
Total	$313.4

Exhibit 16. TOD infrastructure projects by type and cost.
Source: Montgomery County Planning Department, M-NCPPC. *White Flint Sector Plan.* 2010.

---

[60] The commission was empowered by the state of Maryland in 1927 to acquire, develop, maintain, and administer a regional park system in Montgomery and Prince George's counties and to provide land use planning for Prince George's and Montgomery counties.

[61] Montgomery County Planning Department, M-NCPPC. *White Flint Sector Plan.* 2010. p. 73.

FUNDING SOURCES AND FINANCING MECHANISMS

The White Flint Sector Plan is somewhat ambiguous about how public facilities and infrastructure will be financed, stating that "the infrastructure necessary to advance phases of the staging plan should be financed through general fund revenue appropriated in the regular CIP [Capital Improvement Program] process, as well as through mechanisms that would generate significant revenue from properties and developments within the Sector Plan area."[62] Transportation impact fees for new residential and commercial development were already in place when the plan was adopted.

During the planning process, the county explored creating a TIF district that would have diverted a portion of new property tax revenue to fund plan implementation. However, a TIF is unlikely to be implemented in the foreseeable future; it is not clear that the county could make the required finding that redevelopment would not otherwise proceed "but for" the TIF investment, and county officials have expressed concerns about diverting funds that would otherwise flow to the county's general fund.[63]

Several months after approving the plan in 2010, the county council enacted the White Flint Special Tax District, which is authorized to levy a property tax to fund some of the transportation-related infrastructure improvements. The special tax applies to all property in the plan area except for existing residential buildings. The special tax rate was set at $0.103 per $100 of assessed value (for commercial land uses) and is collected with other county property taxes.[64] When the special tax was implemented, the county eliminated the transportation impact fee. Because impact fees depend on new development, they can be an unpredictable revenue source and difficult to finance. By contrast, the special tax spreads the cost over more property owners and provides a more consistent revenue stream against which the county can issue bonds. The county was also concerned that an impact fee would discourage new development.[65]

DENSITY INCENTIVE ZONING PROGRAM

To encourage developers to provide public benefits and to capture some of the value created by upzoning the White Flint area, the county adopted a density incentive zoning program.[66] Under the Commercial Residential Zone that now applies to most of the plan area, developers can choose to either limit their projects to the standard density maximums—0.5 floor area ratio or 10,000 square feet, whichever is greater, and 40 feet in height—or provide public benefits to be allowed to build at higher density. Depending on the area and the level of benefits provided, incentive densities can go up to 3.5 floor area ratio and 300 feet in height. For example, projects can receive incentive density for providing public facilities like schools, libraries, recreation centers, parks, or public parking; providing affordable

[62] Ibid.

[63] Personal communication with Nkosi Yearwood, op cit.

[64] Montgomery County Planning Department, M-NCPPC. *White Flint Sector Plan Implementation Guidelines*. 2011.

[65] Montgomery County Planning Department, M-NCPPC. *White Flint Sector Plan. Appendix 5:Financing*. 2010; Personal communication with Nkosi Yearwood, Area 2 Planner, M-NCPPC by Alison Nemirow, Strategic Economics, on January 27, 2012.

[66] M-NCPPC, Montgomery County Planning Department. *CR Zone Incentive Density Implementation Guidelines*. 2010.

housing beyond the required minimum; building streetscape improvements; or dedicating right-of-way for public use. Projects can also receive extra density for proximity to transit, green building features, and high-quality design.

As of early 2012, two development proposals in the White Flint district were moving into site design. Both take advantage of the density incentives for building public infrastructure such as streetscape improvements and streets.[67]

In addition to the density incentive bonus program, Montgomery County is also considering building on a longstanding parking management program to encourage private developers to build public parking. White Flint is not among the county's four existing parking lot districts but is being considered for a new parking management district (see Exhibit 17 for more information about parking lot districts).

PHASING

The plan includes a three-phase staging plan for transportation infrastructure and real estate development.[68] The phasing will coordinate construction of public facilities with private development to minimize the impacts of traffic congestion and construction on surrounding neighborhoods. It also allows sufficient flexibility so that the county can respond to market forces without losing the plan's vision. The three phases are:

- **Phase 1:** Allows the planning board to approve up to 3,000 dwelling units and 2 million square feet of non-residential development. To move to Phase 2, the county must secure funding for several major street realignments and streetscape projects and achieve a 34 percent non-single-occupancy-vehicle mode share for the plan area (i.e., ensure that at least 34 percent of employees arriving at work in the plan area and residents leaving the plan area during the morning peak period use means other than a single-occupant vehicle, including carpooling, taking transit, walking, or bicycling).

- **Phase 2:** Allows another 3,000 dwelling units and 2 million square feet of non-residential development. To move to Phase 3, the county must begin construction of street realignment and improvement projects. In addition, the county aims to increase the non-single-occupancy-vehicle mode share to 42 percent before moving to Phase 3.

- **Phase 3:** Allows a final increment of 3,800 dwelling units and 1.69 million square feet of commercial development and requires completion of the roadwork and other transportation improvements. The ultimate mode share goal is 51 percent non-single-occupancy-vehicle for residents and 50 percent for employees.

The development cap in each phase is considered officially allocated when developers receive building permits. If the number of units and square feet were considered allocated when projects received initial entitlements, the county would run the risk that large development projects could hoard the available development capacity

---

[67] Personal communication with Nkosi Yearwood, Area 2 Planner, M-NCPPC, by Alison Nemirow, Strategic Economics, on January 27, 2012.

[68] Montgomery County Planning Department, M-NCPPC. *White Flint Sector Plan*. 2010. pp. 67-71.

Exhibit 17. Parking Management in Montgomery County

Montgomery County's parking lot districts are among the earliest examples of a district-based parking management approach in the country. Between 1947 and 1951, Montgomery County established four parking lot districts in its largest central business districts (Bethesda, Montgomery Hills, Silver Spring, and Wheaton). In these districts, developers can choose to pay a special, annual property tax to the county instead of meeting minimum parking requirements on site. The revenue from the tax flows into an enterprise fund in each district and funds public parking construction and operations. Each enterprise fund also receives all public parking revenue collected within the district's boundaries, including revenue from meters, electronic pay stations, cashiered facilities, sale of parking permits, and parking fines. Combined, the four districts provide over 20,000 public spaces. Parking lot district funds have also been used to fund transportation management programs, public transit, and services such as lighting, sidewalks, and streetscape improvements.

In 2009, the Montgomery County Department of Transportation and the M-NCPPC commissioned a study to evaluate the parking lot district program and consider how the county might expand parking management into other commercial centers and corridors such as White Flint. The study concluded that the special property tax, which relies on minimum parking requirements to generate revenue, should be replaced with a tax or fee that either applies to all parking spaces (rather than just those belonging to property owners who choose not to meet minimum parking standards) or is linked to a different metric such as failure to meet mode-split targets or having an "excess" of onsite parking (as measured against a maximum parking standard). Unlinking the districts' funding source from a minimum parking requirement would allow the districts to manage demand rather than providing sufficient supply to meet peak parking demand.

For newly established parking management districts—which might include a new district in White Flint—the study suggests a different approach. The county would establish a target parking range for each land use category, bracketed by minimum and maximum parking requirements based on demand estimates. A developer could provide any quantity of parking spaces within this range and would pay a base annual parking benefit charge that would fund local transportation management programs and, as necessary, public parking. To build above the maximum requirement, the developer could choose to either pay a higher parking benefit charge, to "unbundle" all residential spaces (i.e., separate the price of parking from the rental or sale price of the housing), or "share" (i.e., allow public access to) nonresidential spaces in excess of the maximum. A developer could also elect to provide fewer parking spaces than the required minimum by paying an increased parking benefit charge. In places where parking supply is of particular concern, the study also proposed adjusting each project's parking benefit charge rate as needed to encourage private developers to share their parking supply with the public.

Source: Nelson|Nygaard. *Montgomery County Parking Policy Study: Study Summary*. Prepared for the Montgomery County Department of Transportation and the M-NCPPC. 2011.

As of February 2012, the county was implementing Phase 1 by including transportation improvements in the county's Capital Improvement Program. The special tax district will provide most of the required funding for this phase, and the county has received site plans for 1,000 housing units and 500,000 square feet of commercial development. No building permits have been issued to date under the plan.[69]

---

[69] Personal communication with Nkosi Yearwood, Area 2 Planner, M-NCPPC, by Alison Nemirow, Strategic Economics, on January 27, 2012.

LESSONS LEARNED

The White Flint Sector Plan highlights some of the considerations, including infrastructure financing and phasing, that communities have to take into account to successfully implement TOD, as well as some of the challenges of financing TOD infrastructure.

KEYS TO SUCCESS

- The White Flint planning process involved a diverse group of community and private-sector stakeholders, including a coalition of major property owners and developers. While Montgomery County did not need to hold a public vote to implement the special tax district, support from a broad coalition was critical to ensuring political support for the tax district in the county council.

- Tools like special assessments, TIF, impact fees, and density incentives work by capturing some of the value generated by new development or property value appreciation. Value capture tools like these work well in a place like White Flint because of the district's strong real estate market, which is expected to generated demand for thousands of new residential units and millions of square feet of commercial development.

- The White Flint Sector Plan phasing strategy will help ensure that private land development does not get ahead of public infrastructure, which could lead to traffic congestion or other impacts on surrounding neighborhoods.

KEY BARRIERS

- Implementing the infrastructure and community facilities required to support TOD on the scale envisioned in the White Flint Sector Plan will require years of negotiations with private developers and coordination among multiple public agencies.

- Despite the implementation of a special tax and the density incentive bonus program, Montgomery Council will require other, yet-to-be-determined funding sources to complete all of the infrastructure and community facility projects identified in the plan.

APPLICABILITY TO OTHER PLACES

Financing strategies that use multiple value capture mechanisms over many years to finance districtwide infrastructure and community facilities improvements are most applicable in places where the following conditions are met:

- Strong multiagency and stakeholder coordination and support of a plan.

- A strong real estate market.

- Ability to impose a special tax structure.

REFERENCES

County Council for Montgomery County. Bill 50-10, as adopted November 30, 2010.

County Council for Montgomery County. Resolution No.16-1570, as adopted November 30, 2010.

M-NCPPC, Montgomery County Planning Department. *CR Zone Incentive Density Implementation Guidelines*. 2010. http://www.montgomeryplanning.org/development/documents/CRzoneguidelines.pdf.

Montgomery County Planning Department, M-NCPPC. *White Flint Sector Plan*. 2010. http://www.montgomeryplanning.org/community/whiteflint/documents/WhiteFlintSectorPlanApprovedand Adopted_web.pdf.

Montgomery County Planning Department, M-NCPPC. *White Flint Sector Plan. Appendix 5:Financing*. 2010. http://www.montgomeryplanning.org/viewer.shtm#http://www.montgomeryplanning.org/community/whit eflint/documents/Appendix5.pdf.

Montgomery County Planning Department, M-NCPPC. *White Flint Sector Plan Implementation Guidelines*. 2011. http://www.montgomeryplanning.org/community/whiteflint/documents/WFImplentationGuidelines_web.p df.

Montgomery County Planning Department. "White Flint." http://www.montgomeryplanning.org/community/whiteflint. Accessed January 27, 2012.

Nelson|Nygaard. *Montgomery County Parking Policy Study: Study Summary*. Prepared for the Montgomery County Department of Transportation and the M-NCPPC. 2011. http://www.montgomerycountymd.gov/content/dot/parking/pdf/study_summary.pdf.

Personal communication with Nkosi Yearwood, Area 2 Planner, M-NCPPC, by Alison Nemirow, Strategic Economics, on January 27, 2012.

Ratner, Andrew. "The Upsizing of White Flint." *Planning Magazine*. American Planning Association. October 2011.

G. CORRIDORWIDE TAX-INCREMENT FINANCING: ATLANTA BELTLINE

INTRODUCTION

The Atlanta BeltLine is a $2.8-billion, 25-year redevelopment project financed with a corridorwide tax allocation district (TAD), a mechanism known as TIF in other states.[70] The TAD encompasses approximately 8 percent of the city's area and leverages funding from a wide range of local, state, federal, and philanthropic sources. This case study describes how the TAD works and how the city of Atlanta and its partners plan to combine the TAD with other funding and financing sources to implement the BeltLine plan.

A TAD is a value capture mechanism that depends on property value appreciation and new development to finance infrastructure. However, the advantage of a corridorwide TAD is that the district can use some of the value captured in strong-market parts of the corridor to subsidize infrastructure development in weaker-market areas in the district.

PROJECT BACKGROUND

The Atlanta BeltLine project will knit together segments of four historic rail lines to form a 22-mile transit and greenway corridor circling downtown Atlanta, with transit and greenway spurs connecting 45 different neighborhoods (Exhibit 18). In addition to providing open space and improving transit access, the project is intended to spur private TOD, create jobs, and generally enhance quality of life.

The project was inspired by a Georgia Tech graduate student who proposed a transit system surrounded by a greenway and mixed-use development in his 1999 master's thesis. In 2004, three separate studies concluded that an integrated parks, trail, and transit system was achievable. These studies identified a TAD as a feasible way to pay for 50 to 70 percent of project costs. In 2005, the Atlanta Development Authority worked with other city departments and civic and business leaders to create the BeltLine Redevelopment Plan. As required by state law, the Atlanta City Council, the Fulton County Board of Commissioners,

Exhibit 18. The Atlanta BeltLine TAD (in blue).
Source: Atlanta Development Authority. *Atlanta BeltLine Redevelopment Plan.* 2005.

---

[70] TIF is described in more detail in Appendix B, Section E-3.

and the Atlanta Public School Board of Education approved the redevelopment plan and TAD at the end of 2005. A five-year work plan was created in 2006 to guide the early stages of implementation.[71]

Atlanta BeltLine, Inc. (ABI), a subsidiary of the Atlanta Development Authority, is charged with implementing the plan. The Atlanta BeltLine Partnership, a nonprofit group formed in 2005 to provide community outreach and fund raising, supports ABI's work. The Trust for Public Land and PATH Foundation are involved in land acquisition and development of the greenway. The Metropolitan Atlanta Rapid Transit Authority (MARTA), Fulton County, Atlanta Public Schools, and multiple departments in the city of Atlanta are also collaborating on the project.

Various community partners have been involved in implementation. The Atlanta City Council established five study groups, each charged with master planning efforts for one-fifth of the BeltLine project area. Community representatives also sit on ABI's Board of Directors, the Tax Allocation District Advisory Committee, and the BeltLine Affordable Housing Advisory Board. ABI convenes public briefings on the project's progress four times a year and has established a Community Engagement Advocacy Office that keeps the community informed and manages the study groups.[72]

Specific components of the BeltLine project include:

- Ten master plans covering land use, transportation, and open space for BeltLine subareas.

- Acquisition, planning, and development of 1,280 acres of right-of-way for light-rail or streetcar transit, multiuse trails, and parks.

- Pedestrian improvements, including projects such as traffic calming and streetscape improvements.

- Roadway upgrades, including at-grade crossings, intersection improvements, and new roads linking surrounding neighborhoods to the BeltLine.

- Improvements to existing school facilities and grounds and potentially purchasing land for future schools.

- Infrastructure assistance for private development projects, such as funds for brownfield assessment, environmental remediation, infrastructure construction, and historic preservation.

- Provision of an estimated 5,600 units of affordable housing through the establishment of a $240-million affordable housing trust fund that will provide down-payment assistance to low-income homebuyers and offer grants to finance the acquisition, construction, and/or renovation of affordable multifamily and single-family housing.

---

[71] Atlanta Development Authority. *Atlanta BeltLine Redevelopment Plan*; Atlanta BeltLine. "History of the BeltLine." http://www.beltline.org/BeltLineBasics/BeltLineHistory/tabid/1703/Default.aspx. Accessed September 30, 2011.

[72] Atlanta BeltLine. "Citizen Participation." http://www.beltline.org/BeltLineBasics/CommunityEngagement/Community EngagementOverview/tabid/1829/Default.aspx. Accessed September 30, 2011.

Exhibit 19 shows the estimated cost of each project component, the amount anticipated to be raised by the proceeds of TAD bond sales, and the remaining funds to be raised from other federal, local, and philanthropic sources.

Activity	Total Cost (in millions)	TAD Funds (in millions)	Remaining Need (in millions)
Land acquisition	$480 to $570	$426	$54 to $144
Workforce housing	$220 to $260	$240	--
Greenway design and construction			
BeltLine greenway	$50 to $60	$34	$16 to $26
Connecting greenways	$25 to $30	$19	$6 to $11
Park design and construction	$200 to $250	$120	$80 to $130
Transit construction	$700 to $1,000	$530	$170 to $470
Transportation improvements			
Pedestrian and bicycle projects	$150 to $180	$45	$105 to $135
Operational improvements	$70 to $85	$30	$40 to $55
Atlanta Public Schools projects	$80 to $95	$88	--
Brownfield and intersections	$100	$100	--
Administration and project management	$32	$32	--
Total	**$2,107 to $2,662**	**$1,664**	**$471 to $971**

Exhibit 19. Expected uses of TAD funds and estimated need for other funds.

Source: Atlanta Development Authority. *Atlanta BeltLine Redevelopment Plan* (Table 7.1). 2005.

FUNDING SOURCES AND FINANCING MECHANISMS

The BeltLine TAD encompasses 6,500 acres, or about 8 percent of the city of Atlanta's total land area. In voting to form the TAD in 2005, the city of Atlanta, Fulton County, and Atlanta Public Schools—the three entities that split property taxes generated in this area—agreed to freeze property tax revenue from properties within the TAD at 2005 levels for 25 years. During that period, the TAD will capture any new property tax revenue (known as the "tax increment") generated as development occurs in the district. The Atlanta Development Authority will issue bonds to pay for capital improvements in the district and use the property tax increment to pay back the principal and interest on the bonds.

Georgia state law allows a maximum of 10 percent of the city's tax base to be in a TAD. According to the Atlanta Development Authority, Atlanta has now met that limit (with the BeltLine and other existing TADs) in Atlanta, so no additional TADs can be created.[73]

The BeltLine TAD is projected to raise approximately $1.7 billion over 25 years. Compared to other TADs in Atlanta, the BeltLine TAD will provide limited incentives for private development. The majority of funds will be used to pay for land acquisition, trails, green space, transit and transportation improvements, affordable housing,[74] and Atlanta Public Schools projects. Some funds will be available to

---

[73] Atlanta Development Authority. "Tax Allocation District FAQs," http://www.atlantada.com/buildDev/tadFAQs.jsp#2. Accessed September 30, 2011.

[74] 15 percent of the TAD's net proceeds will be dedicated to the affordable housing trust fund.

encourage private development, primarily for brownfield remediation or to build infrastructure in areas that have been particularly slow to see new development.[75]

In addition to the $1.7 billion raised by TAD bonds, the project will require approximately $1.1 billion in other funds. Exhibit 20 shows additional funding sources identified to date, including a $60 million capital campaign, a variety of local sources, and federal funds.

Sources	Projected Funds
TAD financing	$1.7 billion
Capital campaign*	$60 million
City and local sources[†]	$165 million
Transportation Investment Act[‡]	$602 million
Federal sources:	
Transportation Improvement Program	$18 million
Regional Transportation Plan	$240 million
Other federal funds[§]	$2 million
Total	**$2.8 billion**

Exhibit 20. Expected funding sources.
* $37.5 million had been raised as of 2010.
† City of Atlanta Capital Improvement Program, Park Opportunity Bonds, and Department of Watershed Management funds.
‡ One-percent regional sales tax measure that will go before voters in the 10-county region in 2012.
§ Committed to date for trail construction and transit planning.
Source: Atlanta BeltLine. "How the Atlanta Beltline is Funded." http://beltline.org/about/the-atlanta-beltline-project/funding/. Accessed September 30, 2011.

The capital campaign raised about $37.5 million as of 2010. Approximately $2 million in federal funds have already been committed for trail construction and transit planning. The Atlanta Regional Commission, the region's MPO, has planned for $18 million in federal funds to be used for right-of-way acquisition and trail construction in the region's Transportation Improvement Program and included $240 million in federal funds for transit development in the Regional Transportation Plan. Finally, the Atlanta Regional Roundtable—a county commission charged with selecting a list of transportation projects to be included in an upcoming regional sales tax proposal—has included $602 million for BeltLine light-rail projects on the list. The Transportation Investment Act, which would establish a 1-percent sales tax, will go before voters in the 10-county region in July 2012.[76]

As of early 2012, seven of the 10 required master plans had been adopted, with the remainder nearing completion, and the environmental impact statement for the transit component of the project was underway. The Atlanta Development Authority had begun raising funds from TAD bond sales, and ABI and the Atlanta BeltLine partnership had started to assemble funding from private donations and state,

---

[75] Atlanta BeltLine. "Tax Allocation District (TAD)." http://www.beltline.org/Funding/TaxAllocationDistrictTAD/tabid/1731/Default.aspx. Accessed September 2011; Atlanta Development Authority. *Atlanta BeltLine Redevelopment Plan.* 2005.

[76] Atlanta BeltLine. "Transportation Investment Act." http://www.beltline.org/Funding/TransportationInvestmentAct/tabid/4138/Default.aspx. Accessed October 2, 2011.

federal, and other sources. The affordable housing trust fund was capitalized with $8 million from early bond sales. Finally, ABI had acquired 280 acres of green space, completed three parks and begun construction on five others, and opened several segments of multiuse and hiking trails.[77]

LESSONS LEARNED

KEYS TO SUCCESS

- A broad coalition is required to implement a project on the scale of the Atlanta BeltLine. Support from community members, nonprofits, the private sector, multiple government agencies and departments, and civic leaders has been instrumental to advancing the project from a dream to a concrete plan and to the funding and financing strategy. Because Georgia state law (like many other states) requires all affected taxing entities to vote to establish a TAD, support from the city of Atlanta, Fulton County, and Atlanta Public Schools was especially critical to making the financing plan work.

- The TAD or TIF district model relies heavily on property tax increment created by new development and therefore requires a strong real estate market to be viable. However, by extending the district to the entire corridor, the Atlanta BeltLine model creates the opportunity to fund infrastructure projects in relatively weak-market subareas in the corridor using the revenue generated in stronger-market subareas.

KEY BARRIERS

- A corridorwide infrastructure financing strategy involves multiple neighborhood groups and many different property owners and therefore requires extensive community outreach and education. In the case of the Atlanta BeltLine, this has included 10 different master planning processes, multiple implementation committees, and quarterly progress briefings. Regardless of geographic scope, implementing a TIF district or TAD is often a lengthy process that, depending on state law, can involve extensive public process and negotiations among multiple entities.

- In addition to the TAD, the BeltLine project will require a wide range of other funding sources, including other local funds, federal grants, public and private donations, and potentially revenue from a regionwide sales tax.

APPLICABILITY TO OTHER PLACES

A corridor-level TAD or TIF district can be useful for places where public infrastructure needs to be extensively rebuilt and there is potential to use the value captured in stronger-market subareas to support weaker-market subareas in the corridor. A corridor-level district might have the best chance of succeeding politically in a place when the entire transit corridor falls within the boundaries of one jurisdiction, limiting the number of taxing entities that must be involved in forming the TIF district or

---

[77] Sheperd, J. et al. "The Atlanta BeltLine—A Model of Urban Transformation." Presented at the 10th Annual New Partners for Smart Growth conference, February 3, 2011.

TAD. Indeed, even if the entire proposed district is within one city, the process can still be complicated by the involvement of multiple agencies.

REFERENCES

Atlanta BeltLine. "Home page." http://beltline.org. Accessed October 31, 2011.

Atlanta BeltLine. *Annual Report.* 2010. http://beltline.org/Portals/26/Atlanta%20Beltline%20AnnualReport-web%20file.pdf.

Atlanta BeltLine. "Citizen Participation." http://www.beltline.org/BeltLineBasics/CommunityEngagement/Community EngagementOverview/tabid/1829/Default.aspx. Accessed September 30, 2011.

Atlanta BeltLine. "History of the BeltLine." http://www.beltline.org/BeltLineBasics/BeltLineHistory/tabid/1703/Default.aspx. Accessed September 30, 2011.

Atlanta BeltLine. "How the Atlanta Beltline is Funded." http://beltline.org/about/the-atlanta-beltline-project/funding/. Accessed September 30, 2011.

Atlanta BeltLine. "Tax Allocation District (TAD)." http://www.beltline.org/Funding/TaxAllocationDistrictTAD/tabid/1731/Default.aspx. Accessed September 2011.

Atlanta BeltLine. "Transportation Investment Act." http://www.beltline.org/Funding/TransportationInvestmentAct/tabid/4138/Default.aspx. Accessed October 2, 2011.

Atlanta Development Authority. *Atlanta BeltLine Redevelopment Plan.* 2005. http://www.beltline.org/LinkClick.aspx?fileticket=5d1Fx%2bLu6ic%3d&tabid=1820&mid=3489.

Atlanta Development Authority. *Plan of Work for 2006-2010 Budget.* 2006. http://www.beltline.org/Portals/26/Media/PDF/Final%20WorkPlan20July05.pdf.

Atlanta Development Authority. "Tax Allocation District FAQs," http://www.atlantada.com/buildDev/tadFAQs.jsp#2. Accessed September 30, 2011.

Sheperd, Joyce; Leary, Brian; Wilson, Valarie; and Conable, Nate. "The Atlanta BeltLine—A Model of Urban Transformation." Presented at the 10th Annual New Partners for Smart Growth conference. February 3, 2011. http://www.newpartners.org/2011/docs/presentations/thurs/NP11_Anderson_Sheperd_Leary_Wilson_Con able.pdf.

H. MULTISTATION TAX-INCREMENT FINANCING: DALLAS TOD TIF DISTRICT

INTRODUCTION

TIF allows the public sector to "capture" the value of growth that results from new development and increasing property values. That increment of growth funds public improvements to revitalize the TIF district. While TIF is a powerful tool, it can present a "chicken-and-egg" problem in areas that require public improvements to unlock the potential for development: TIF revenue accrues after property values go up, but revenue is needed to fund public improvements required to spur new development and increased property values.[78]

A corridorwide or multistation TIF district helps address the chicken-and-egg problem by capitalizing on increases in property values in one area to make improvements in another area. This type of TIF district is an especially appealing alternative along a transit corridor, where real estate market conditions and community needs can vary greatly among different station areas. For example, one station area in a multistation TIF district might have a stronger real estate market and therefore more immediate potential for growth in value, while another station area might have a weaker market and need more investment to build potential for new development.

In general, value capture mechanisms such as TIF are easier to implement within single jurisdictions because they typically rely on a local tax or fee. They can be challenging to use on a transit corridor that crosses jurisdictional boundaries. The city of Dallas implemented a multistation TIF district in cooperation with several overlapping jurisdictions. Compared to the Atlanta BeltLine TAD (discussed in the previous case study), which covers 6,500 acres and will finance a new transit line in addition to TOD infrastructure, the Dallas district is smaller (1,167 acres) and will contribute to TOD infrastructure but will not finance a new transit line.

PROJECT BACKGROUND

In 2008 the city of Dallas approved a transit-oriented development tax increment financing district (TOD TIF District) along a Dallas Area Rapid Transit (DART) light rail corridor. The process of planning, developing new policies, and conducting negotiations between the city and multiple partners and stakeholder groups, including DART, Southern Methodist University, and a local real estate firm, to establish the district took four years. As originally approved, the TOD TIF District covered 558 acres.[79] In 2010, the TOD TIF District was expanded to include 1,167 acres in four subdistricts (shown in Exhibit 21):

- Mockingbird/Lovers Lane Subdistrict.

- Cedars West Subdistrict.

- Cedar Crest Subdistrict.

- Lancaster Corridor Subdistrict.

---

[78] TIF is described in more detail in Appendix B, Section E-3.

[79] Kitty and Michael Dukakis Center for Urban and Regional Policy, Northeastern University. "Corridor-Based Tax Increment Financing Districts." http://www.dukakiscenter.org/tif-districts. Accessed October 2011.

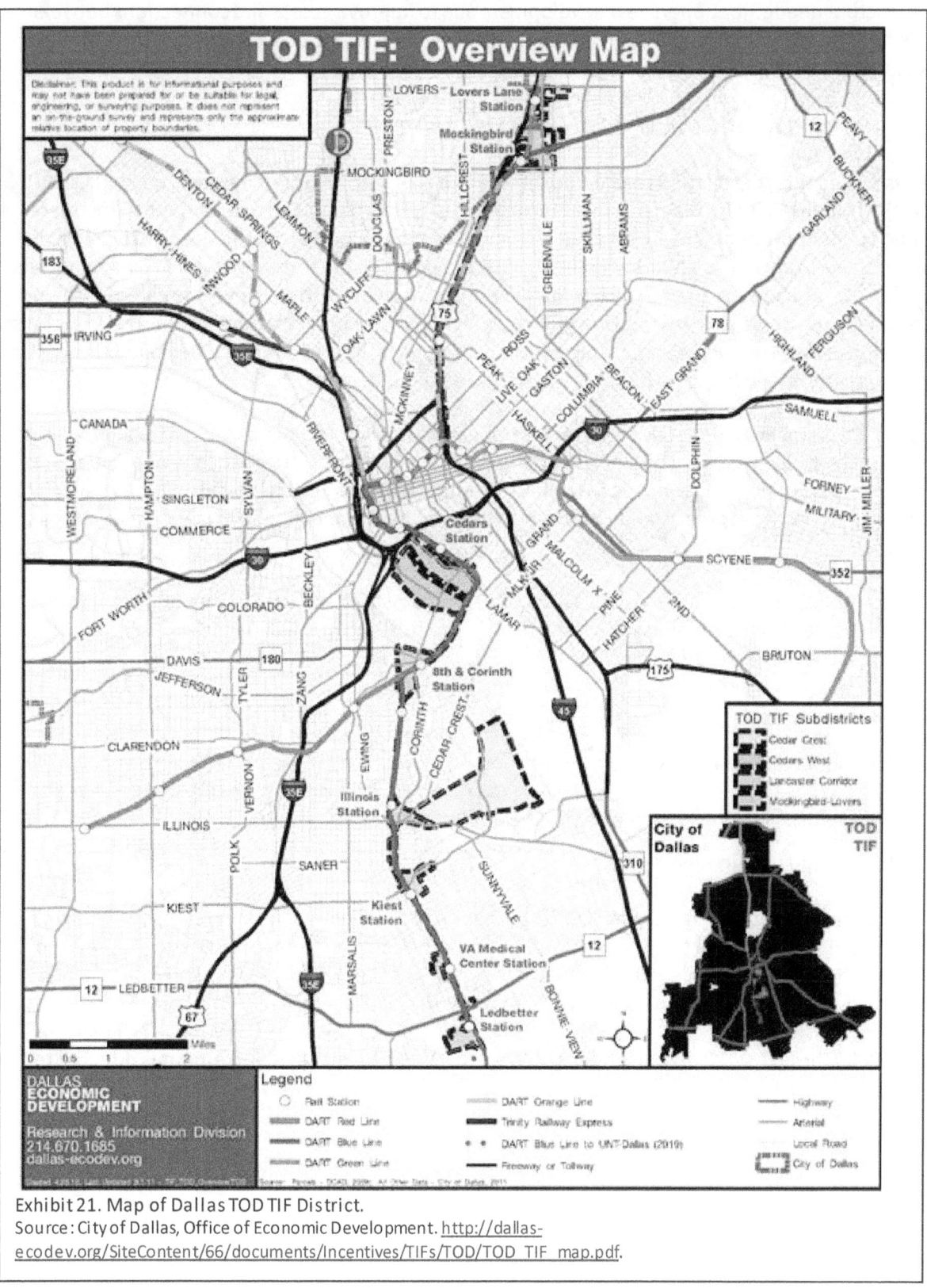

Exhibit 21. Map of Dallas TOD TIF District.
Source: City of Dallas, Office of Economic Development. http://dallas-ecodev.org/SiteContent/66/documents/Incentives/TIFs/TOD/TOD_TIF_map.pdf.

A primary purpose of the TOD TIF District is to encourage high-density, mixed-use, walkable station areas along the existing DART line. To that end the TIF revenue is being used to pay for the public infrastructure needed to support new development and to improve access and connections between the existing DART station areas and surrounding institutional uses, including Southern Methodist University, the George W. Bush Presidential Library, the Trinity River, and Veterans Memorial Hospital.

FUNDING SOURCES AND FINANCING MECHANISMS

Over its 30-year life, the district is projected to generate over $185 million in tax increment (in 2009 dollars). The Dallas TOD TIF District was created to stimulate development in certain subdistricts by funding public improvements with revenue generated by other subdistricts. The Dallas TOD TIF District allows revenue from the neighborhoods in the northern portion of the corridor, which have higher land values and greater potential for growth in the increment, to be used in less-developed areas in the Lancaster Corridor area south of the Trinity River, which has more infrastructure needs. The TIF will also provide infrastructure and pedestrian improvements around DART stations that would not otherwise be possible, as well as funding for affordable housing throughout the district.[80]

In Dallas, TIF districts are project-specific: a developer for a new project typically pays for and/or constructs the project area's public infrastructure improvements and is subsequently reimbursed from TIF revenue as property values increase and tax increment revenue becomes available. This project-driven, pay-as-you-go approach limits the financial risk for the city and other participating government agencies.

In Texas, multiple entities can choose to participate in a TIF district, and the amount of increment dedicated to the TIF is negotiated and agreed to district by district. In the Dallas TOD TIF, the other entities that overlap in the district include:

- Dallas County.

- Dallas Independent School District.

- Dallas County Community College District.

- Dallas County Hospital District.

The level of participation in the Dallas TOD TIF for each entity varies (and in some cases varies over the term of the TIF), but none of the entities are directing all of the tax increment they would otherwise receive to the TIF District. For example, the city of Dallas' participation in the TOD TIF District is for a 30-year period, and the level of participation follows a modified bell curve:

- For 2009 through 2011, 70 percent of the city's portion of generated increment is directed to the TOD TIF District.

---

[80] City of Dallas Office of Economic Development. "Dallas TOD Experience and TOD TIF District: Providing Unique Public Financial Incentives for Transit Oriented Development in Underserved Areas." July 16, 2010.

- For 2012 through 2029, 85 percent of the city's portion of generated increment is directed to the TOD TIF District.

- For 2030 through 2038, 70 percent of the city's portion of generated increment is directed to the TOD TIF District.

Dallas County's level of participation in the TOD TIF District is set at 55 percent of generated increment from 2011 through 2030. After 2030, the county will retain all of the property tax revenue it is due.[81]

The Dallas TOD TIF funds projects case by case rather than having a list of planned infrastructure improvements for the district. Types of TOD infrastructure to be funded by the increment include:

- Public infrastructure, including water, wastewater, stormwater, paving, streetscape, and utility burial or relocation.

- Environmental remediation and demolition.

- Parks, open space, and trails.

- Façade restoration.

- Transit-related improvements.

- Affordable housing.[82]

In addition to funding infrastructure, the increment can be used for grants to help finance TOD projects in the district. The TIF revenue will be used for the infrastructure improvements needed for individual development projects and to improve pedestrian connections to DART stations from the surrounding neighborhoods. See Exhibit 22 for an example of a project receiving TIF revenue from the TOD TIF District.

Exhibit 22. Dallas TOD TIF District Project Example: Lancaster Urban Village Project

The planned Lancaster Urban Village project in the Lancaster Subdistrict is a development project receiving tax increment funding for TOD infrastructure. The mixed-use project will include 193 residential units, 20 percent of which are required to be affordable; onsite amenities including a clubhouse and swimming pool; 14,000 square feet of retail and small office space; and structured parking to serve the project and an adjacent 46,568-square-foot expansion of the Dallas Urban League, a job-training and social service nonprofit agency. A groundbreaking ceremony occurred in March 2012. The following table shows funding sources and uses for the $25.8 million project.

Funding Source	Amount
HUD (221(d)(4) loan)	$12,400,000
City of Dallas Section 108	$7,400,000
Public-private partnership	$3,200,000
New markets tax credits	$2,800,000
Total	$25,800,000

The project is receiving a TIF contribution that will repay the city's Section 108 loan and partially repay the public-private partnership funding. The TIF contribution will be used for infrastructure improvements as shown in the following table.

TIF Improvement Category	Amount
Infrastructure	$2,200,000
Demolition	$300,000
Grant for high-density project	$1,700,000
Affordable housing	$4,300,000
Total	$8,500,000

The infrastructure costs to be funded include stormwater upgrades in the Lancaster Corridor Subdistrict that must be addressed before any redevelopment can occur in the area. No other existing city funds can pay for the proposed stormwater upgrades.

Source: City of Dallas Office of Economic Development. "Dallas TOD Experience and TOD TIF District." 2010.

---

[81] City of Dallas Office of Economic Development. *TOD TIF District Plan*. 2010. p. 63.

[82] Ibid, p. 59.

A fundamental goal of the Dallas TOD TIF District is to permit tax increment sharing from the Mockingbird/Lovers Lane Subdistrict to trigger redevelopment of the Lancaster Corridor Subdistrict in the city's southern sector, where development has lagged for many years. The financing plan for the TOD TIF District allocates 40 percent of the increment generated from the Mockingbird/Lovers Lane Subdistrict to the Lancaster Corridor Subdistrict. An additional 20 percent of the increment from the Mockingbird/Lovers Lane Subdistrict will be allocated to the districtwide affordable housing budget. The remaining 40 percent of the Mockingbird/Lovers Lane Subdistrict increment will be used for projects in that area.[83]

The financing plan allocates 10 percent of the increment generated in the Cedars West Subdistrict to the Lancaster Corridor Subdistrict and 10 percent to the districtwide affordable housing budget. The remaining 80 percent of the Cedars West Subdistrict increment will be used for projects in that subdistrict because Cedars West has significant infrastructure needs.

The Cedar Crest Subdistrict increment will be largely retained for projects in that area, although a small portion will go to the districtwide affordable housing budget. Exhibit 23 shows the projected amounts of tax revenue increment and how it is allocated among the subdistricts.

Subdistrict / Category	Estimate of Value From New Development	2009 Tax Revenue Increment Generated	2009 Budget Allocation
Lancaster Corridor	$171,000,000	$13,200,000	$49,780,000
Cedar Crest	$326,000,000	$25,830,000	$25,330,000
Mockingbird/Lovers Lane	$840,000,000	$76,560,000	$30,020,000
Cedars West	$1,094,000,000	$69,590,000	$54,580,000
Affordable housing (all subdistricts)	N/A	N/A	$21,830,000
Administration and implementation	N/A	N/A	$3,640,000
Total	$ 2,260,000,000	$185,180,000	$185,180,000

Exhibit 23. Estimated increment generation and allocation by subdistrict.
Source: City of Dallas Office of Economic Development. *TOD TIF District Plan*. 2010. p. 58.

This increment-sharing arrangement means that the subdistrict with the highest needs, Lancaster Corridor, will be allocated increment exceeding the amount of revenue it generates. As shown in Exhibit 23, the Lancaster Corridor sub-district is projected to generate $13 million in increment, but will be allocated almost $50 million for public infrastructure and other improvements.

Each subdistrict is allocated some increment, but the amount varies based on the improvements needed to stimulate revitalization and improve pedestrian connections rather than strictly on the amount of increment generated in the subdistrict itself. This type of increment sharing allows the district to use market momentum in one area to stimulate development in another. The affordable housing funds generated will be available districtwide to help projects fulfill some of the city and county's affordable housing requirements.[84]

---

[83] Ibid, p. 57.

[84] City of Dallas Office of Economic Development. *TOD TIF District Plan*. 2010. pp. 60 and 64.

As of late 2011, a mixed-use project, with 55 dwelling units and 3,720 square feet of retail valued at $9.5 million, has been completed. In addition to the Lancaster Urban Village Project described in Exhibit 22, several significant development projects with a total projected value of about $85 million have been planned or are under construction in the TOD TIF District, including two mixed-use projects and a boutique hotel. Southern Methodist University, located in the Mockingbird/Lovers Lane Subdistrict, is also planning major projects in the area, including the George W. Bush Presidential Library and campus facilities.

LESSONS LEARNED

KEYS TO SUCCESS

- TIF revenue can be applied to infrastructure that does not generate revenue, making it applicable to a wide variety of infrastructure.

- A multistation TIF district relies heavily on property tax increment from new development and therefore requires a strong real estate market to be viable. As with the Atlanta BeltLine, the Dallas TOD TIF District creates the opportunity to fund infrastructure projects in relatively weak-market subareas using the revenue generated in stronger-market subareas.

KEY BARRIERS

- Implementation for a multistation TIF district tends to be lengthy because it requires negotiations with multiple stakeholder groups, which can include overlapping jurisdictions, neighborhood groups, and property owners. In the Dallas TOD TIF District, the planning and negotiation process took four years.

- TIF is a cross-subsidy from some public services to others, and public authorities need to understand such trade-offs. Because the property taxes allocated to other services are frozen for a long time, inflationary pressures and population growth tend to quickly diminish their per capita value, affecting the quality and quantity of the services funded through property taxes.

- Even with some development activity occurring in the Dallas TOD TIF District in 2010, the total assessed value in the district declined due to the overall economic conditions and decline in property values. The decline in property values means that the district is not meeting its projections for tax revenue generated. Such shortfalls could result in TOD infrastructure projects being delayed.

- Despite the clear benefits of using TIF across multiple station areas and allowing revenue generated in one station area to be deployed in another, the city of Dallas anticipates that TIF revenue alone will be insufficient to cover all of the costs for TOD infrastructure improvements in the district. Even with the potential to share increment across subdistricts, the city anticipates pursuing a diverse set of other potential funding and financing sources, including:[85]

---

[85] City of Dallas Office of Economic Development. "Dallas TOD Experience and TOD TIF District: Providing Unique Public Financial Incentives for Transit Oriented Development in Underserved Areas." July 16, 2010.

- City of Dallas sources, including the Office of Economic Development, the Housing Finance Corporation, and general obligation bonds.

- Dallas County's Capital Improvement Program.

- North Central Texas Council of Governments sustainable development grants.

- New Market, Low-Income Housing, historic preservation, and other tax credits.

- DART Surplus Property Program.

- HUD CDBG Section 108 Loan Program (mezzanine loan funding).

- Special assessment districts (called public improvement districts in Texas).

APPLICABILITY TO OTHER PLACES

Laws regulating the use of TIF vary from state to state, so the applicability of a multistation TIF district would also vary, but a TIF district is an especially appealing alternative along a transit corridor where real estate market conditions and community needs can vary greatly among the different station areas. This district-level financing tool presents a particular opportunity for places where public infrastructure needs to be rebuilt and where the value captured from strong-market subareas could fund improvements in weaker-market subareas. Because TIF, like most value capture mechanisms, is designed to be deployed within a single jurisdiction, corridors or districts that encompass multiple jurisdictions are likely to be more difficult locations to employ a TIF strategy.

REFERENCES

City of Dallas Office of Economic Development. *TOD TIF District Plan*. 2010. http://www.dallas-ecodev.org/SiteContent/66/documents/Incentives/TIFs/TOD/TOD_TIF_plan.pdf.

City of Dallas Office of Economic Development. "Transit Oriented Development (TOD) Tax Increment Financing (TIF) District." http://www.dallas-ecodev.org/business/tifs/todTIF.htm. Accessed October 24, 2011.

City of Dallas Office of Economic Development. "Dallas TOD Experience and TOD TIF District: Providing Unique Public Financial Incentives for Transit Oriented Development in Underserved Areas." July 16, 2010. http://www.dallas-ecodev.org/SiteContent/66/documents/Incentives/TIFs/Dallas%20TIF%20Overview_TOD%20TIF_BetterHouston%20Group_7-16-10.pdf.

Kitty and Michael Dukakis Center for Urban and Regional Policy, Northeastern University. "Corridor-Based Tax Increment Financing Districts." http://www.dukakiscenter.org/tif-districts. Accessed October 24, 2011.

I. SUPPORTING TOD WITH FEDERAL TRANSPORTATION GRANTS: TRANSPORTATION FOR LIVABLE COMMUNITIES

INTRODUCTION

The Bay Area Metropolitan Transportation Commission's Transportation for Livable Communities (TLC) grant program funds projects that support TOD, including streetscape improvements, non-transportation infrastructure, transportation demand management projects, and land banking or site assembly. The TLC program has allowed the Metropolitan Transportation Commission (MTC), the region's MPO, to use state and federal transportation funds (including CMAQ and Transportation Enhancement funds)[86] creatively to support compact housing and mixed-use projects close to transit.

This type of grant program, which directs federal and sometimes state transportation funding to support TOD, is usually implemented at the regional level by an MPO, which allocates most state and federal transportation funds in metropolitan areas. The ability to create such a program depends on the level of discretion that the state legislature and department of transportation allow MPOs in allocating state and federal transportation funds, as well as on the willingness of the MPO's board members and other regional stakeholders to prioritize TOD infrastructure over other types of transportation improvements.

PROGRAM BACKGROUND

The TLC program grew out of MTC's Transportation/Land-Use Connection policy, adopted in 1996 to better coordinate regional transportation and land-use planning.[87] The TLC program was launched in 1997 to direct transportation improvement funds to support local governments' infill, TOD, and neighborhood revitalization efforts. The program originally consisted of three components:

- The TLC Planning Program (created in 1997) funded community planning efforts to revitalize existing neighborhoods, downtowns, commercial cores, and transit stops and create more pedestrian-, bicycle-, and transit-friendly environments.

- The TLC Capital Program (created in 1998) funded transportation infrastructure improvements that encourage pedestrian, bicycle and transit trips and support compact, mixed-use development.

- The Housing Incentives Program (created in 2000) rewarded communities with funding for TLC-type transportation improvements when they built compact housing and mixed-use developments at transit stops.[88]

Between the late 1990s and 2007, funding for the TLC program expanded from an original annual commitment of $9 million to $27 million a year. Major funding sources included the federal CMAQ

---

[86] The CMAQ and Transportation Enhancement programs are discussed in Appendix B, Sections F-1 and F-2, respectively.

[87] MTC functions as both the regional transportation planning agency—a state designation—and, for federal purposes, as the region's MPO. A separate council of governments, the Association of Bay Area Governments, is charged with regional land use planning in the Bay Area.

[88] MTC. *Ten Years of TLC: An Evaluation of MTC's Transportation for Livable Communities Program.* 2008. p.3.

Program and the Surface Transportation Program (STP), including transportation enhancement funding. One-third of the funding was allocated to the TLC Planning and Capital Programs, another third to the Housing Incentives Program, and the final third to the region's nine county Congestion Management Agencies for local TLC Capital and/or Housing Incentives programs. During this period, MTC designated $84 million for 81 capital projects, with the average grant size ranging from $600,000 in the early funding cycles to over $1.5 million in later cycles. Typical projects funded by the Capital Program included bicycle routes, transit access improvements, and pedestrian facilities such as improved sidewalks, crosswalks, lighting, and streetscape amenities (Exhibit 24).[89]

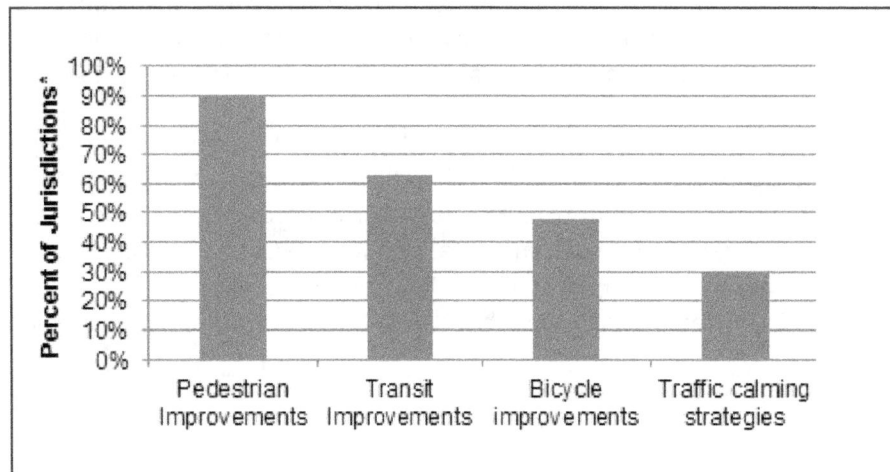

Exhibit 24. Types of transportation improvements funded by TLC Capital Program. *Based on survey of 56 project sponsors awarded funding between 1998 and 2005. Source: MTC. *Ten Years of LTC: An Evaluation of MTC's Transportation for Livable Communities Program*. 2008.

In 2007 and 2008, MTC conducted an internal evaluation of the program[90] and commissioned a white paper from the Center for Transit-Oriented Development[91] that examined options for strengthening MTC's programs to foster development that revitalizes central cities and older suburbs, supports and enhances public transit, promotes walking and bicycling, and preserves open spaces and agricultural lands. Recommendations from the two evaluations included more directly supporting infill housing and TOD by focusing on designated priority development areas;[92] replacing the TLC Planning Program with larger land use planning grants and smaller technical assistance grants; discontinuing the Housing Incentives Program; providing larger Capital Program grants at more frequent intervals; and broadening eligible projects for the Capital Program to include parking, land assembly, and non-transportation infrastructure. Based on these recommendations, in 2010 MTC reconfigured its planning grants, eliminated the Housing Incentives Program, and added three new categories of eligible capital projects beyond their traditional focus on streetscape projects. The new project categories were:

---

[89] Ibid. p.4.

[90] MTC. *Ten Years of TLC: An Evaluation of MTC's Transportation for Livable Communities Program*. 2008.

[91] Center for Transit-Oriented Development. *Financing Transit-Oriented Development in the San Francisco Bay Area: Policy Options and Strategies*. 2008.

[92] Priority development areas are locally identified, infill development opportunity areas in existing communities that MTC has designated to accommodate the majority of future regional population and employment growth.

- Transportation demand management (TDM) projects that result in more efficient use of transportation resources, such as those incorporating Clipper (the region's electronic transit fare card), carsharing, or parking management strategies.

- Non-transportation infrastructure improvements, such as sewer upgrades.

- Direct TOD funding for land banking or site assembly.[93]

The TDM category can, in theory, include parking garages. However, MTC requires project sponsors who request funds for parking garages to complete cost-benefit analyses for their projects and demonstrate that TDM options cannot sufficiently reduce parking demand. When MTC added the new project categories, it also implemented new scoring criteria that place more emphasis on TDM, affordable housing, and project readiness; increased the maximum grant size; increased the required local match from 11.5 to 20 percent; and required that projects be located in priority development areas.

After the program changes, MTC issued a call for projects (in 2010), funding 23 projects at an average of $1.9 million each. Because of complications involved in funding the non-transportation infrastructure improvements and land assembly project categories (discussed below), only two projects in these categories received funding: $2.5 million for land acquisition for an affordable housing project in the city of Livermore and $1.045 million for water and sewer supply upgrades in the city of Santa Rosa. The other projects included the typical range of pedestrian, bicycle, and transit access improvements.

FUNDING SOURCES AND FINANCING MECHANISMS

Exhibit 25 shows the breakdown of funding sources that MTC used for the TLC program between 1996 and 2009.[94] The CMAQ program is the main source for the TLC Capital Program, though MTC has also sometimes used federal Transportation Enhancement funds. Federal STP funds have been used primarily for planning activities. Early rounds of the program were funded partially with state Transportation Development Act money.

STP, CMAQ, and Transportation Enhancement funds can be applied directly to bicycle, pedestrian, and

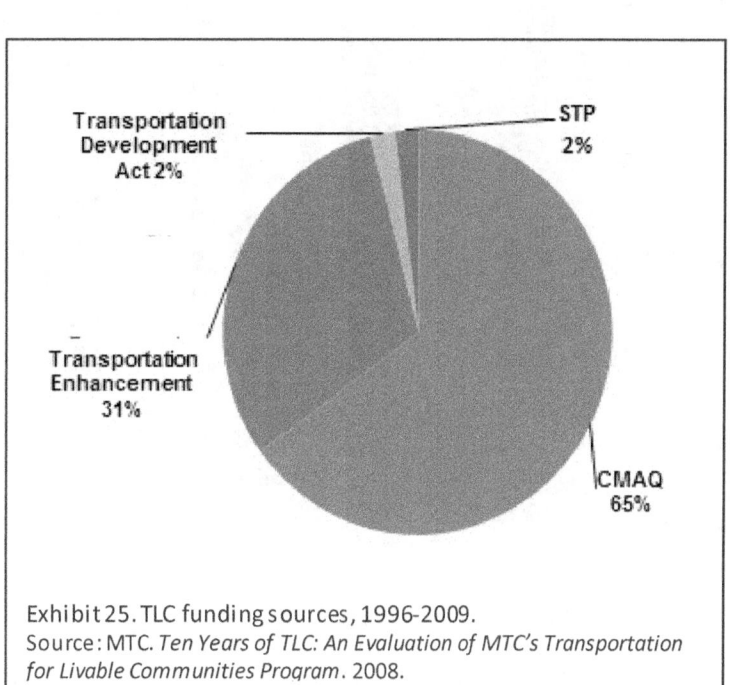

Exhibit 25. TLC funding sources, 1996-2009.
Source: MTC. *Ten Years of TLC: An Evaluation of MTC's Transportation for Livable Communities Program*. 2008.

---

[93] MTC. "Transportation for Livable Communities Program." http://www.mtc.ca.gov/planning/smart_growth/tlc. Accessed September 30, 2011; MTC. "Memorandum: Proposed Transportation for Livable Communities (TLC) Goals and Scoring Criteria." December 31, 2009.

[94] The most recent year for which data are available.

streetscape improvements. These sources cannot, however, directly fund non-transportation infrastructure or land assembly, two of the TLC project categories added in 2010. To use TLC funds for these purposes, cities must be prepared to exchange CMAQ funding from the TLC program with other discretionary dollars that have been designated to fund a local transportation project. The local sources that are freed up by MTC's CMAQ funding can then pay for the non-transportation infrastructure or land assembly TLC project. [95]

TLC is intended to leverage other funding and financing sources rather than to cover the entire cost of a project. Federal guidelines require an 11.5-percent local match for the funding sources that MTC uses for the TLC program. To ensure that localities are committed to projects, MTC increased the local match requirement to 20 percent in 2010.

The 2007 internal evaluation found that the local match averaged 76 percent, although the matches likely included funding for other components of a larger project. [96] The evaluation also found that the TLC funds were often some of the earliest funding, helping project proponents attract other funding for later stages of the project. Exhibit 26 shows the other types of funding sources that are typically invested in TLC projects.

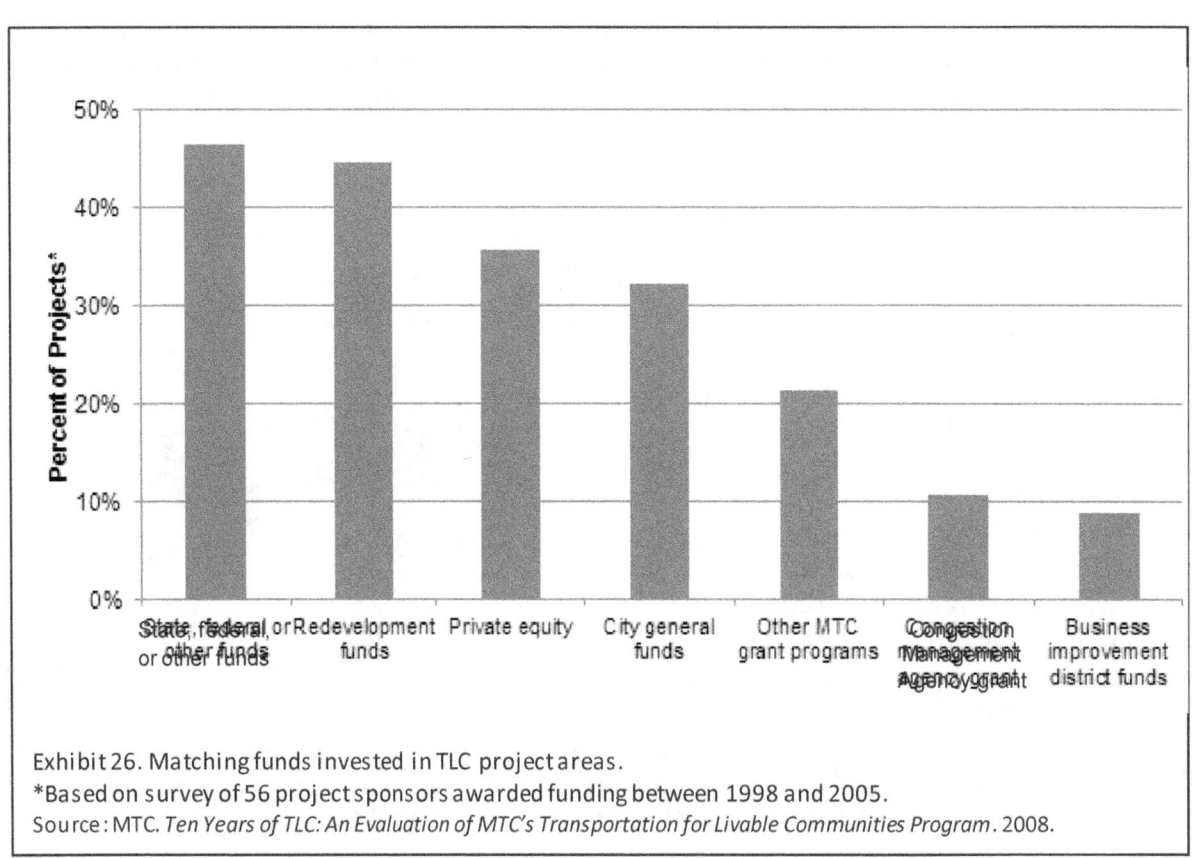

Exhibit 26. Matching funds invested in TLC project areas.
*Based on survey of 56 project sponsors awarded funding between 1998 and 2005.
Source: MTC. *Ten Years of TLC: An Evaluation of MTC's Transportation for Livable Communities Program.* 2008.

---

[95] MTC has also used another model for this type of fund exchange; in 2010, it contributed $10 million to the Bay Area Transit-Oriented Affordable Housing Fund that was funded by exchanging CMAQ and Transportation Enhancements money for discretionary funds from one of the Congestion Management Agencies. See Chapter III, Section J, for more information.

[96] MTC. *Ten Years of TLC: An Evaluation of MTC's Transportation for Livable Communities Program.* 2008. p. 13.

In May 2012, MTC established a new "OneBayArea Grant Program" that shifts more of the agency's discretionary federal funding—including the funding for the TLC Capital Program—to the county Congestion Management Agencies. A 1990 California ballot measure requires CMAs coordinate transportation planning, funding, and other congestion management activities in each county as d by. Congestion Management Agencies have significant discretion over how to use the funds, although they are limited by CMAQ eligibility requirements[97] and by a requirement that 50-70 percent of all funds be spent in priority development areas, depending on the size of the county. Congestion Management Agencies are required to adopt a PDA Investment and Growth Strategy that focuses transportation investments on infill development and plans for affordable housing development.[98]

LESSONS LEARNED

The TLC program directs federal transportation funds, such as CMAQ and STP, to streetscape and other transportation enhancements that support land use planning goals. The program also allows MTC to fund non-transportation activities, like land assembly and sewer upgrades, by swapping federal transportation dollars for local funds. The program's success in attracting more applications than it is able to fund attests to local jurisdictions' need for funding. Other MPOs, including Portland Metro in Oregon and the North Central Texas Council of Governments in Dallas-Fort Worth, have offered similar programs that swap federal transportation dollars with local funds to support land assembly for TOD.[99]

KEYS TO SUCCESS

- California's method of distributing CMAQ and STP funds provides MTC with a significant amount of flexibility. MTC has taken advantage of this flexibility to funnel CMAQ funds in particular to TLC and other bicycle and pedestrian grant programs.

- MTC created the TLC program to support community-based transportation projects that bring new vibrancy to downtown areas, commercial cores, neighborhoods, and transit corridors. This mission has guided TLC's evolution over time.

KEY BARRIERS

- MTC has found that the level of grant funding the TLC program can provide is not adequate to encourage residential and mixed-use development. This finding led MTC to eliminate the Housing Incentive Program in 2010. Rather than creating incentives for new development, the TLC program fills funding gaps and improves bicycle and pedestrian connections from transit stations to surrounding neighborhoods.

---

[97] CMAQ would make up approximately half of the funds that each Congestion Management Agency would receive.

[98] MTC "OneBayArea Grant Program." http://www.mtc.ca.gov/funding/onebayarea/. Accessed July 23, 2012.

[99] Center for Transit-Oriented Development. *Financing Transit-Oriented Development in the San Francisco Bay Area: Policy Options and Strategies.* 2008.

- For an MPO to use federal transportation dollars for non-transportation infrastructure or land assembly, localities must be able and willing to exchange local funds for CMAQ or STP funds. This exchange requires significant effort from a local government.

APPLICABILITY TO OTHER PLACES

This model could be implemented in states that allow MPOs discretion in allocating state and federal transportation funds. MPOs that identify TOD as a funding priority would need to devote staff time and money to creating a TOD funding program.

REFERENCES

Center for Transit-Oriented Development. *Financing Transit-Oriented Development in the San Francisco Bay Area: Policy Options and Strategies.*
2008. http://www.mtc.ca.gov/planning/smart_growth/tod/Financing_TOD_in_SFBA.pdf.

MTC. "Memorandum: Proposed Transportation for Livable Communities (TLC) Goals and Scoring Criteria." December 31,
2009. http://apps.mtc.ca.gov/meeting_packet_documents/agenda_1424/3_TLC_Scoring_Criteria.pdf.

MTC "OneBayArea Grant Program." http://www.mtc.ca.gov/funding/onebayarea/. Accessed October 5, 2011.

MTC. *Ten Years of TLC: An Evaluation of MTC's Transportation for Livable Communities Program.*
2008. http://www.mtc.ca.gov/planning/smart_growth/tlc/tlc_eval/10_Years_of_TLC_Eval_Summary_2008.pdf.

MTC. "Transportation for Livable Communities Program." http://www.mtc.ca.gov/planning/smart_growth/tlc. Accessed September 30, 2011.

Personal communication with Therese Trivedi, Transportation Planner, MTC, by Alison Nemirow, Strategic Economics, on October 5, 2011.

J. STRUCTURED FUND FOR TOD LAND ACQUISITION: BAY AREA TRANSIT-ORIENTED AFFORDABLE HOUSING ACQUISITION FUND

INTRODUCTION

The San Francisco Bay Area Transit-Oriented Affordable Housing (TOAH) Acquisition Fund is a $50-million structured fund[100] that provides financing for acquiring land for affordable housing development near transit. Structured funds are a kind of loan fund that pools money from different investors with varying expectations of risk and return for a dedicated purpose.

When it was launched in 2011, the Bay Area TOAH Fund represented the state of the art in TOD structured funds. While other structured funds, such as the Denver TOD Fund (discussed in Exhibit 28), provide financing to acquire and/or develop affordable housing near transit, the TOAH Fund is unique among structured funds dedicated to TOD because it operates at the regional level. Moreover, the fund's structure builds on the experiences of other funds with similar goals.

PROGRAM BACKGROUND

In 2008, recognizing that the downturn of the housing market and financial recession created an opportunity for preserving property near transit for permanent affordable housing, the Great Communities Collaborative (GCC), a Bay Area partnership of national and regional advocacy, research, and funding organizations dedicated to promoting affordable housing and TOD,[101] began discussing the creation of a fund to acquire property for affordable TOD. GCC commissioned the Center for Transit-Oriented Development to conduct a feasibility study, which recommended forming a short-term structured loan fund modeled after existing funds developed by Enterprise Community Partners, Inc., and the Low Income Investment Fund (LIIF) in other parts of the country. The feasibility study also recommended tailoring the fund to overcome specific barriers to equitable TOD in the Bay Area, including a scarcity of development sites near transit, relatively high land costs, and the difficulty of acquiring property before securing project financing.[102] GCC convened a steering committee composed of representatives from the Metropolitan Transportation Commission (MTC), the region's MPO; the Association of Bay Area Governments, the region's council of governments; the affordable housing development community; and the core GCC partners to clarify the fund's goals and how it would operate.

---

[100] Structured funds are discussed in more detail in Appendix C. Fundamentals of Structured Funds.

[101] GCC partners include the Greenbelt Alliance, the Nonprofit Housing Association of Northern California, Transform, Urban Habitat, Reconnecting America, the San Francisco Foundation, the Silicon Valley Community Foundation, and the East Bay Community Foundation. For more information, see http://www.greatcommunities.org.

[102] Center for Transit-Oriented Development and Strategic Economics. *San Francisco Bay Area Property Acquisition Fund for Equitable Transit-Oriented Development: Feasibility Assessment Report.* Prepared for the Great Communities Collaborative. 2010.

In 2010, GCC approached MTC about investing through the Transportation for Livable Communities Program. [103] MTC's board committed $10 million to the fund by exchanging CMAQ and STP money for discretionary funds from one of the region's county Congestion Management Agencies.

MTC's commitment was critical because it served as a top-loss investment, meaning that any defaults would affect MTC's investment first, reducing the risk to other potential investors. With this upfront commitment, GCC and the San Francisco Foundation released a request for proposals from prospective fund managers. In July 2010, LIIF and a consortium of five other community development financial institutions (CDFIs) were selected, with LIIF as the fund manager and administrative agent. LIIF and its CDFI partners created a business model for the fund, which determined that its maximum size could be $50 million based on the available top-loss capital, the underwriting criteria, and loan interest rates. The CDFI consortium was also responsible for raising additional capital for the fund beyond the top-loss contribution from MTC. Based on the fund structure, it quickly attracted money from a variety of sources, including capital from the CDFIs themselves, program-related investments[104] from foundations, and bank loans. The fund raised the maximum it could support with the $10 million in top-loss money, and in March 2011 the final documents were executed and loans became available.

FUNDING SOURCES AND FINANCING MECHANISMS

Exhibit 27 illustrates how the TOAH Fund is structured. The $10-million investment from MTC occupies the top-loss risk position in the fund. LIIF and its partners raised $15 million from six CDFIs, the Ford Foundation, the San Francisco Foundation, and Living Cities.[105] The investment from the CDFIs and foundations occupies the second tier. Two banks, with a cumulative investment of $25 million, occupy the most secure, senior risk position.[106]

If any of the fund's loans default, MTC will take the first loss, followed by the CDFIs and foundations, and finally the banks. In other words, each layer in the stack "protects" the next layer. The relatively high risk tolerance and no- or low-interest expectations of the MTC, foundations, and CDFIs, combined with the protection that the structure provides for the banks' investment, allows the TOAH Fund to offer loans with low interest rates and high loan-to-value ratios compared to comparable commercial products. The fund is intended to exist for 10 years. During the first five years, it will originate loans; in the final five years, it will only collect repayment and will not make any additional loans.

The Bay Area TOAH Fund offers five type of loans for affordable housing, community facilities, and neighborhood services, including:

---

[103] The TLC program, established in the late 1990s, provides assistance for capital projects—typically bicycle, pedestrian, and transit access or streetscape projects -- that support TOD. See Chapter III, Section 9, for more information on TLC.

[104] More information about program-related investments is in Appendix B, Section F-7.

[105] Living Cities is a philanthropic collaborative of 22 foundations and financial institutions dedicated to improving the lives of low-income people and the cities where they live.

[106] If the debtor goes bankrupt, investors in the senior risk position must be repaid before other creditors.

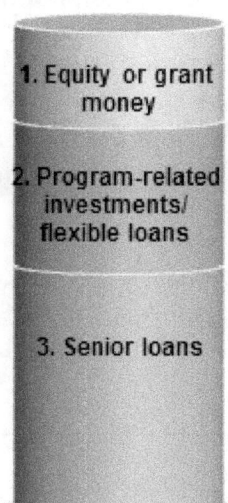

1. Equity or grant money

Top loss (public sector): $10 million from MTC

2. Program-related investments/ flexible loans

Second loss (foundations and CDFIs): $15 million from six CDFIs and Ford Foundation, San Francisco Foundation, and Living Cities

3. Senior loans

Third loss (banks): $25 million from Morgan Stanley and Citi Community Capital

Exhibit 27. Bay Area TOAH Fund structure.
Source: Bay Area TOAH. "Bay Area's Transit-Oriented Affordable Housing Fund." Presented at Affordable Housing Week. May 2011.

- **Predevelopment loans** for costs incurred in predevelopment, including architecture, engineering, environmental studies, surveys, market studies, appraisals, hazard and liability insurance, property taxes, site security, financing fees, and debt service expenses.

- **Acquisition loans** to acquire vacant land or operating housing or commercial property and to cover lot development expenses.

- **Construction bridge loans** to bridge the time period between construction funding and either larger or longer-term financing.

- **Construction-to-mini-permanent loans** for construction financing (new or rehabilitation) followed by a small permanent loan to pay off the short-term construction loan.

- **Leveraged loans** to fund eligible predevelopment, acquisition, construction, and/or mini-permanent financing to leverage an investment into a new market tax credit-eligible transaction, which could be community facilities, neighborhood retail, fresh food markets, child care centers, or similar facilities. [107]

With the exception of predevelopment loans, which are capped at $750,000, the maximum loan size is $7.5 million. The loan-to-value ratio for most of the products can go up to 110 percent. Eighty-five percent of the fund's capital is targeted to create and preserve affordable housing. Fifteen percent can be used to support community facilities, health clinics, fresh food markets, and other neighborhood retail projects. Projects must be within a half-mile of high-quality transit service, which includes BART,

---

[107] Bay Area TOAH. "Projects." http://bayareatod.com/projects. Accessed September 28, 2011.

light rail, and bus rapid transit, and in one of the priority development areas designated by the Bay Area's regional agencies to accommodate the majority of future population and employment growth.

The TOAH Fund's target borrowers are experienced nonprofit or for-profit developers, municipal agencies, and joint ventures with strong track records in affordable housing development. Developers can apply to any of the six participating CDFIs for a loan. The originating CDFI underwrites the loan request and submits it to LIIF, which prepares a package for the fund's credit committee to review. If the loan is approved, the originating CDFI services the loan throughout repayment.[108]

As of late 2011, the TOAH Fund has loaned $3 million for an affordable housing project in San Jose and $7.3 million for an affordable housing project with a ground-floor grocery store in San Francisco.[109] Exhibit 28 describes another TOD fund that helps to develop and preserve affordable housing.

Exhibit 28. The Denver Transit-Oriented Development Fund

The Denver TOD Fund, established in 2010, is a $15-million fund that aims to develop and preserve 1,200 affordable housing units near transit over 10 years. In contrast to TOAH, which is a stand-alone entity, the Denver TOD fund is operated by the Enterprise Community Loan Fund. The Denver TOD Fund includes $2.5 million in top-loss funds from the city of Denver, comprised of $2 million from the city's Excel Energy franchise fee revenue and $500,000 in economic development business incentives funds; $1 million in second-loss funds from Enterprise Community Partners; and $4.5 million in third-loss funds from various foundations. The Enterprise Community Loan Fund and the Mile High Community Loan Fund invested $5.5 million in senior debt. The Urban Land Conservancy also contributed a $1.5-million equity investment.

The conservancy is the only entity approved to borrow from the fund. It contributes 10 percent of the equity to every project. It works with prospective affordable housing developers to identify opportunities, takes out a short-term loan from the fund to purchase the site and its properties, and eventually sells or leases the property to the developer when permanent financing becomes available to pay back the loan. Enterprise and other loan investors agreed to make the conservancy the sole borrower.

Because the Denver TOD Fund received top-loss funds only from the city, it cannot finance projects outside city boundaries. The fund is also limited by the region's reliance on federal Low Income Housing Tax Credits for permanent financing; because the city of Denver can expect only two Low Income Housing Tax Credit projects annually, the fund cannot have more than two loans that expire in any given year. Another significant difference from TOAH is that part of the Urban Land Conservancy's mission is to purchase and hold opportunity sites in corridors that are slated for future transit development. Because vacant land typically does not generate revenue, the conservancy could have difficulty finding sites that can make interest payments.

Source: Center for Transit-Oriented Development and Strategic Economics. *CDFIs and Transit-Oriented Development.* Appendix B. Prepared for the Federal Reserve Bank of San Francisco. 2010.

---

[108] Bay Area TOAH. "Bay Area's Transit-Oriented Affordable Housing Fund." Presented at Affordable Housing Week. May 2011.

[109] Bay Area TOAH. "Projects." http://bayareatod.com/projects. Accessed September 28, 2011.

LESSONS LEARNED

Funds must be tailored to fit a locality or region's need and to take advantage of the financial resources available. Different TOD needs require different loan products. For example, affordable housing preservation requires larger loans than land acquisition. The scale and availability of public-sector support and foundations' investments, the number and sophistication of CDFIs and community development corporations, enthusiasm from banks, and the availability of take-out financing[110] will affect the size and structure of the fund that a region can support. [111]

KEYS TO SUCCESS

- The many gatherings and studies that went into establishing the TOAH Fund ensured that the fund was tailored to meet the Bay Area's needs. The process of bringing stakeholders together, preparing a compelling case for a structured fund, and identifying an appropriate source of top-loss money provided national investors, including national foundations, CDFIs, and banks, with the confidence that the fund would be viable and well structured.

- Public and foundation investments that occupy the top-loss risk positions are critical to the success of structured funds. To meet the critical financing gap, a fund must typically offer loans with some combination of low interest rates, high loan-to-value ratios, longer terms, larger loan amounts, and/or softer recourse requirements. [112] Public investments and foundation program-related investments can absorb the risk of default, and their below-market return expectations allow funds to offer better terms. In the case of TOAH, MTC's $10-million top-loss investment was the critical piece that allowed the fund to attract other investors. Because senior debt is by definition conservative in the loss risk that lenders are willing to accept, the size of MTC's investment and the second-loss position investments largely determined the overall size of the fund and its allowable loan-to-value ratio. MTC's top-loss investment could be spread across the region, which made it possible to form a regional fund. Other funds, such as Denver's, have been unable to find top-loss risk position investments at the regional level, limiting the funds' scope to the jurisdiction that is willing and able to occupy the top-loss risk position.

KEY BARRIERS

- Establishing TOAH took approximately three years; Denver's TOD Fund took closer to four. Identifying the housing financing need, making the case for the fund and attracting investors, finding a fund manager, and negotiating the optimal fund structure all take time. [113]

---

[110] Take-out financing is an agreement by a lender to pay off a construction loan and leave the developer with permanent, long-term financing when construction is finished.

[111] Center for Transit-Oriented Development and Strategic Economics. *CDFIs and Transit-Oriented Development.* Appendix B. Prepared for the Federal Reserve Bank of San Francisco. 2010.

[112] Recourse requirements are obligations the borrower agrees to if unable to pay the debt.

[113] Center for Transit-Oriented Development and Strategic Economics, *CDFIs and Transit-Oriented Development.* Appendix B. Prepared for the Federal Reserve Bank of San Francisco. 2010.

- Because structured funds require repayment and are ultimately accountable to investors, they can only fund activities that generate revenue and/or can anticipate receiving permanent financing in a given period of time.

APPLICABILITY TO OTHER PLACES

To create a structured fund, local and/or regional governments need to work with foundations and investors to identify a specific financing need, prepare a compelling case for a structured fund, identify a source of top-loss money, and negotiate an appropriate fund structure. This process is critical to give national foundations, CDFIs, banks, and other potential investors the confidence that the community can create a viable fund.

Identifying top-loss money is easiest in states like California that allow regional and local transportation authorities significant discretion in allocating federal transportation funds.

REFERENCES

Bay Area TOAH. "Home page." http://bayareatod.com. Accessed September 30, 2011.

Bay Area TOAH. "Projects." http://bayareatod.com/projects. Accessed September 28, 2011.

Bay Area TOAH. "Bay Area's Transit-Oriented Affordable Housing Fund." Presented at Affordable Housing Week. May 2011. http://bayareatod.com/wp-content/uploads/2011/05/TOAH-Fund-Presentation-05-10-11.pdf.

Center for Transit-Oriented Development and Strategic Economics. *CDFIs and Transit-Oriented Development*. Prepared for the Federal Reserve Bank of San Francisco. 2010. http://www.frbsf.org/publications/community/wpapers/2010/cdfi_transit_oriented_design.html.

Center for Transit-Oriented Development and Strategic Economics. *San Francisco Bay Area Property Acquisition Fund for Equitable Transit-Oriented Development: Feasibility Assessment Report*. Prepared for the Great Communities Collaborative. 2010. http://bayareatod.com/wp-content/uploads/2011/11/Bay_Area_TOD_Fund_DrftRprt_SE11310.pdf.

City of Denver, Office of Strategic Partnerships. "Denver Transit Oriented Development Fund." http://www.denvergov.org/DenverOfficeofStrategicPartnerships/Partnerships/DenverTransitOrientedDevelopmentFund/tabid/436574/Default.aspx. Accessed January 30, 2012.

Great Communities Collaborative. "Home page." http://www.greatcommunities.org. Accessed January 30, 2012.

K. REGIONAL TOD INVESTMENT FRAMEWORK: CENTRAL CORRIDOR LIGHT RAIL AND THE CENTRAL CORRIDOR FUNDERS COLLABORATIVE

INTRODUCTION

The Central Corridor is an 11-mile light-rail corridor planned to run between downtown St. Paul and Minneapolis, Minnesota (see Exhibit 29). After many years of project planning, construction on the alignment began in 2010; service is expected to begin in 2014. The project planning focused not only on the light rail construction, but also on stimulating TOD along the corridor. The Central Corridor passes through several disinvested neighborhoods that could benefit from the infrastructure investment, change in land uses, and boost in property values that TOD can provide. The local governments, along with an active nonprofit community, has developed tools to prime the area for future development, focusing on property acquisition and infrastructure development near the planned transit stations. The Central Corridor light rail demonstrates the importance of regional coordination and cooperation to create a defined vision for TOD investment.

PROJECT BACKGROUND

Local planning for transit along the corridor began as early as 1981, involving numerous stakeholders in the process. In 2006, the Metropolitan Council, the Twin Cities' MPO, submitted a plan for light rail to the Federal Transit Administration (FTA) and sought permission to initiate preliminary engineering.[114] FTA gave approval in December 2006. Between early 2007 and the start of construction in 2010, the

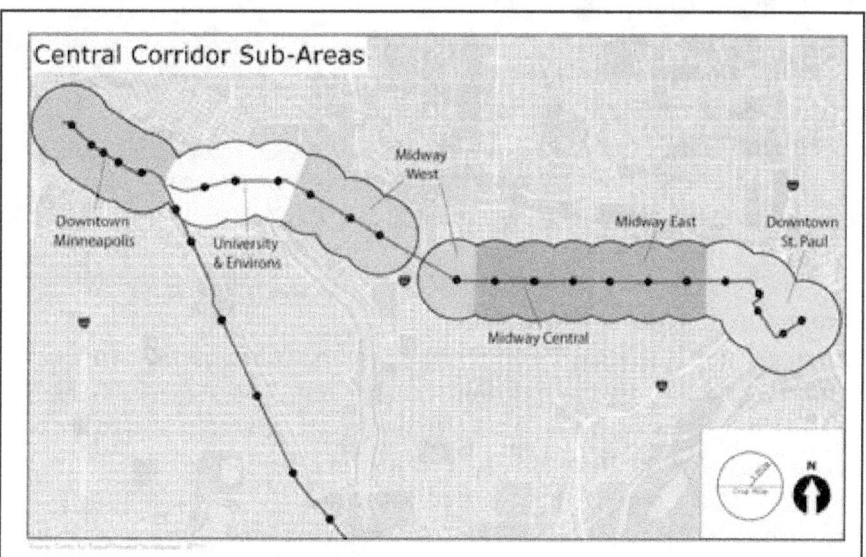

Exhibit 29 Central Corridor subareas.
Source: Center for Transit-Oriented Development. *Central Corridor Investment Framework: A Corridor Implementation Strategy.* 2010.

Metropolitan Council and project partners finalized station locations, determined project costs, and, most importantly, committed local funds to the project.[115] FTA New Starts funding requires local agencies demonstrate that they can fund, implement, and operate major transit projects.

---

[114] FTA's New Starts program funds major transit capital investments. Projects approved for the New Starts program must be evaluated throughout the entire project development process.
[115] Metropolitan Council. "Central Corridor Light Rail Transit." http://www.metrocouncil.org/transportation/ccorridor/centralcorridor.asp. Accessed October 24, 2011.

The Metropolitan Council secured funding guarantees from local and state agencies, including the state of Minnesota; Hennepin and Ramsey counties; the city of St. Paul; and the newly formed Counties Transit Improvement Board. In 2011, FTA committed to pay half the cost of construction, which is $957 million as of 2012 (see Exhibit 30).

The light rail will improve access to five major economic activity centers: the Minneapolis and St. Paul downtowns, the Midway commercial district, the Minnesota State Capitol complex, and the University of Minnesota. These economic centers contain almost 280,000 jobs and are projected to have 345,000 by 2030.[116] The Central Corridor light rail will connect to other regional public transit, giving people who live along the corridor access to more job opportunities and making the jobs in the Central Corridor easier to reach from elsewhere in the Twin Cities region.

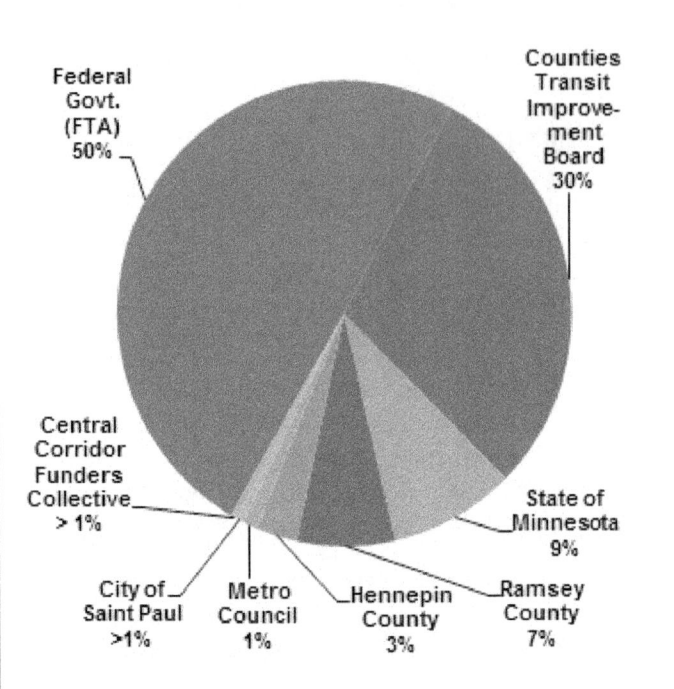

Exhibit 30. Funding sources for Central Corridor light rail.
Source: Metropolitan Council. "Central Corridor LRT: Frequently Asked Questions."
http://www.metrocouncil.org/transportation/ccorridor/ccfaq.htm. Accessed October 24, 2011.

In summer 2009, the Central Corridor Working Group (CCWG) was formed to develop a coordinated investment framework for TOD along the Central Corridor. The working group included representatives from the Metropolitan Council, Hennepin and Ramsey counties, the cities of Minneapolis and Saint Paul, and the Minnesota Housing Finance Agency. CCWG, with the assistance of outside consultants, identified more than 500 projects in the areas around the planned light-rail stations, the first step in creating and prioritizing a comprehensive list of future TOD investments. The projects included streets and sidewalks, bikeways, streetscapes, public art, parks, housing, and office, retail, and hotel space.

The potential public improvements that CCWG identified totaled over $6 billion, and CCWG has sought to leverage private investment to help pay for some of these costs. CCWG believed that a relatively small public-sector infrastructure investment in these areas could raise nearby property values and accelerate private investment. Exhibit 31 shows potential infrastructure investment needs by subarea.

---

[116] Ibid.

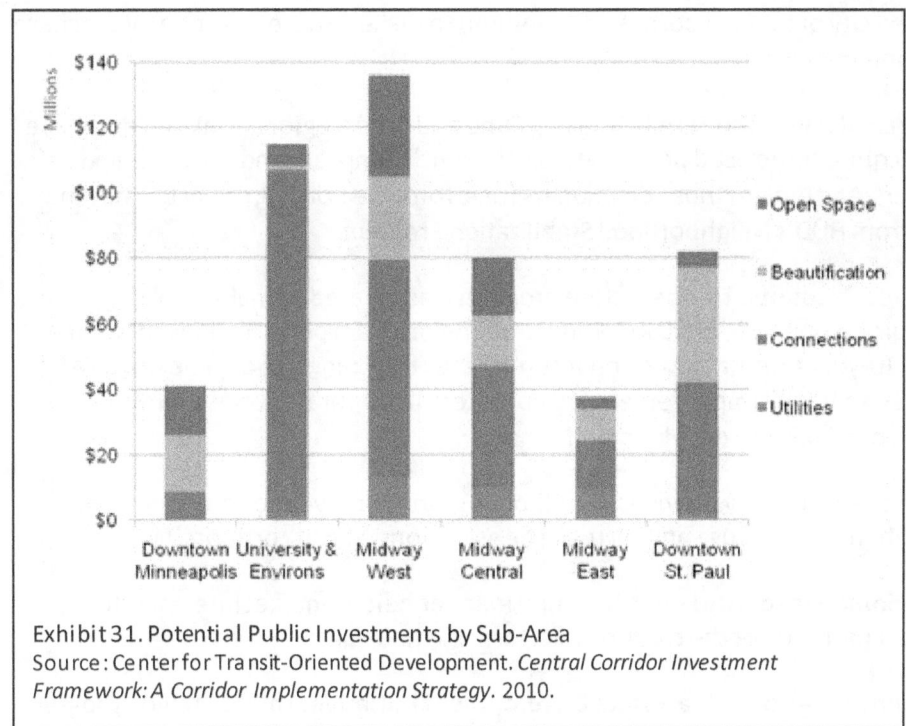

Exhibit 31. Potential Public Investments by Sub-Area
Source: Center for Transit-Oriented Development. *Central Corridor Investment Framework: A Corridor Implementation Strategy*. 2010.

FUNDING SOURCES AND FINANCING MECHANISMS

The corridor has nearly 800 acres of underutilized land valued at over $1 billion that could be used for TOD.[117] Public and philanthropic funds are most likely to meet the need for medium- to long-term, flexible, patient capital to acquire properties.[118] Efforts to assemble underused land and catalyze infrastructure investment along the Central Corridor drew the interest of the nonprofit and philanthropic communities. The Central Corridor Funders Collaborative (CCFC) is a partnership of 12 local and national philanthropic organizations formed to help catalyze change along the new rail line. CCFC promotes affordable housing, a strong local economy, vibrant transit-oriented places, and effective communication and collaboration. CCFC created a Catalyst Fund through which it plans to invest $20 million in the corridor over 10 years. As of late 2011, the group has raised $5 million to invest in corridorwide strategies and efforts.[119] Collaboration between CCWG and CCFC helps ensure that public and nonprofit entities are working together to encourage TOD along the Central Corridor.

In addition to the Catalyst Fund, other funds supporting TOD along the Central Corridor include:

- **Land Acquisition for Affordable New Development Fund:** Minnesota Housing, the Metropolitan Council, and the Family Housing Fund (a community development corporation) collaborated to create an $11-million pilot fund to support land acquisition by cities, community development corporations, or housing authorities with preference given to projects near transit. The fund is intended to support mid-term project-level investments. The acquired parcels cannot have ready-to-go projects, and funds must be spent within one year and repaid within five years. Any appreciation in the value of land acquired through the program can be rolled into the project to support affordable housing, and any losses in land value will be covered by the fund. A pilot loan program

---

[117] Center for Transit-Oriented Development. *Property Acquisition for TOD in the Central Corridor: Acquisition Fund Framework Study Final Reports*. 2009.

[118] Center for Transit-Oriented Development. *Central Corridor Investment Framework: A Corridor Implementation Strategy*. December 2010.

[119] CCFC. "About Us." http://www.funderscollaborative.org/about-us. Accessed October 24, 2011.

started in 2009, when the city of St. Paul borrowed $2 million to make a strategic property purchase along the light-rail alignment.

- **Twin Cities Community Land Bank:** The Family Housing Fund and other regional stakeholders have formed a land bank to acquire foreclosed properties, partner with nonprofit and socially-minded for-profit housing developers, and lend to those developers for affordable housing projects. The land bank received funding from HUD's Neighborhood Stabilization Program.

- **Transit Improvement Area Accounts:** This new state program was created to make public improvements and acquire property for TOD in Minnesota. The program plans to allow loans of up to $2 million with up to 10-year terms at low or no interest rates for a range of eligible uses. To be eligible, an area must have a transit improvement area plan that incorporates transit with commercial, residential, or mixed-use development.

- **County Bond Funds:** Hennepin County provides $2 million in grants each year on a two-year cycle for TOD projects that enhance transit use and increase density along transit corridors.

- **Family Housing Fund's Home Prosperity Fund:** This fund loans at below-market interest rates to community development partners for the creation of affordable housing.

- **Neighborhood Development Center's Real Estate Development Initiative:** This $1-million program is designed to give entrepreneurs business training and help buying commercial property. The Neighborhood Development Center has collaborated with community development corporations and has partnered with the Community Reinvestment Fund[120] to develop a standard loan package for the program.

- **Local Initiatives Support Corporation Acquisition and Predevelopment Funds:** The Local Initiatives Support Corporation supports nonprofit developers by offering short-term acquisition loans and predevelopment recoverable grants that provide money for expenses incurred before permanent construction financing is secured. The grants are repaid at 0% interest from construction or permanent financing proceeds. The amount of funding and terms vary annually.[121]

LESSONS LEARNED

The Central Corridor highlights the importance of establishing a framework for TOD investment through a collaborative network of community entities early in the project development.

KEYS TO SUCCESS

- Coordination and cooperation among government agencies and the philanthropic community to develop a cohesive vision for the Central Corridor before the light rail was built encouraged investment in the area.

---

[120] The Community Reinvestment Fund is a national nonprofit that provides capital to nonprofit community development lenders through its secondary market for loans.

[121] Center for Transit-Oriented Development. *Property Acquisition for TOD in the Central Corridor: Acquisition Fund Framework Study Final Reports.* 2009.

- The CCWG provided the framework for effective collaboration to create a more defined vision for TOD investment in the Central Corridor. The effort relied on many groups because the corridor crosses several jurisdictional boundaries (Hennepin County, Ramsey County, and the cities of Minneapolis and St. Paul) as well as connecting several neighborhoods and anchor institutions.

KEY BARRIERS

- Redeveloping along the Central Corridor poses challenges because much of the land adjacent to the light-rail alignment is in a weak market. The area needs to be primed for private-sector development before TOD projects can be built. However, in this market context, particularly during the economic downturn, the private-sector might be unwilling or unable to invest resources that could result in bigger changes. The philanthropic community needs help from other entities, and current resources might be insufficient to meet the needs.

- Many of the funding programs discussed in this case study have not started any concrete projects, in part because they are new, but also because of the weak real estate market. As stations are built, the market could gain momentum to implement TOD.

APPLICABILITY TO OTHER PLACES

This model for a regional TOD investment framework could be applicable to other places if public infrastructure needs to be rebuilt across multiple jurisdictions. Success would likely depend on the presence of a strong and engaged philanthropic community in addition to the participation of the local governments, including MPOs or other regional agencies, housing authorities, and redevelopment agencies. The participation of nonprofit community development corporations whose organizational goals coincide with TOD principles can help this model succeed.

REFERENCES

Center for Transit-Oriented Development. *Central Corridor Investment Framework: A Corridor Implementation Strategy.* 2010. http://www.funderscollaborative.org/sites/default/files/pdfs/Central_Corridor_Investment_Framework_report.pdf.

Center for Transit-Oriented Development and Strategic Economics. *CDFIs and Transit-Oriented Development.* Prepared for the Federal Reserve Bank of San Francisco. 2010. http://www.frbsf.org/publications/community/wpapers/2010/cdfi_transit_oriented_design.html.

Center for Transit-Oriented Development. *Property Acquisition for TOD in the Central Corridor: Acquisition Fund Framework Study Final Reports.* 2009. http://www.funderscollaborative.org/sites/default/files/Central%20Corridor%20Land%20Fund-Report-Final.pdf.

CCFC. "About Us." http://www.funderscollaborative.org/about-us. Accessed October 24, 2011.

Metropolitan Council. "Central Corridor Light Rail Transit." http://www.metrocouncil.org/transportation/ccorridor/centralcorridor.asp. Accessed October 24, 2011.

Metropolitan Council. "Central Corridor LRT: Frequently Asked Questions." http://www.metrocouncil.org/transportation/ccorridor/ccfaq.htm. Accessed October 24, 2011.

IV. INNOVATIVE MODELS

This chapter explores four emerging, innovative models for meeting infrastructure needs for TOD:

- Anchor institution partnerships.

- Corridor-level parking management.

- Land banking.

- District energy systems.

The models were chosen because they address some of the most pressing challenges of providing TOD infrastructure, including paying for infrastructure in weak real estate markets, financing structured parking facilities, acquiring and assembling land for TOD, and capitalizing on efficiencies in the construction process. Unlike some of the more traditional infrastructure financing tools, implementing these models is not as simple as developing a project proposal and applying for a grant or issuing debt. Rather, the models are long-term strategies that involve building partnerships among multiple jurisdictions and institutions, thinking strategically about how to prioritize resources throughout a corridor or region, and looking for synergies among different infrastructure projects.

Each model description includes a discussion of:

- The model's role in funding and financing TOD infrastructure.

- Case studies that illustrate how the model might work in practice or, for models that have never been fully tested, examples of how different components of the model have been implemented.

- Lessons learned for implementing the model, including the types of places where the model might work and what entities would be involved in implementation.

A. ANCHOR INSTITUTION PARTNERSHIPS

Over the last two decades, local governments and advocates for neighborhood revitalization have begun to urge anchor institutions—nonprofit or private entities such as universities, hospitals, and corporations that are inextricably tied to their locations because of real estate holdings, capital investment, history, or mission—to orient their development decisions and day-to-day operations around improving the economic health of surrounding neighborhoods, including by encouraging transit use and TOD. Anchor institutions could facilitate infrastructure development by providing upfront funding for planning and design, convening community leaders and other stakeholders around common goals, and catalyzing economic reinvestment that could enhance the tax base. Involving institutions in local community and economic development could work wherever anchor organizations are present, but it is particularly relevant in weak-market places where nonprofit institutions are often the largest employers and could have the most incentive to invest in their communities.

To help local governments and organizations engage anchor institutions, this section discusses:

- The role of anchor institution strategies in funding and financing TOD infrastructure and the advantages for local governments in involving them.

- A case study of University Circle, a district in Cleveland, Ohio, that is home to approximately 40 education, health, arts, and social service organizations, which demonstrates various ways in which institutions can work with public agencies to address critical infrastructure needs.

- Lessons learned and how this model might apply to other places.

THE ROLE OF ANCHOR INSTITUTION PARTNERSHIPS IN FUNDING AND FINANCING TOD INFRASTRUCTURE

Historically, the classic example of an anchor institution was a local bank or other hometown corporation that invested money locally, hired local employees, and otherwise contributed to the success of the local and/or regional economy. Although many companies still invest in their communities, as the economy has globalized and corporations have become increasingly mobile, advocates for mobilizing anchor institution investments have shifted their attention over the last decade or two towards major nonprofit organizations such as universities, colleges, health centers, museums, libraries, and performing arts centers. These institutions are unlikely to relocate and therefore have an interest in the success of their surrounding communities. Moreover, universities, colleges, and hospitals now rank among the largest private employers in many cities. For example, the University of Pennsylvania is the largest private employer in Philadelphia, and Johns Hopkins Medical Center is the largest private employer in Maryland. [122]

Anchor institutions have used many strategies to invest in their communities. [123] Examples include:

---

[122] Webber, H. S. and Karlström, M. *Why Community Investment is Good for Nonprofit Anchor Institutions*. Chapin Hall at the University of Chicago. 2009. p. 6.

[123] Ibid.

- **Live local, buy local, hire local:** As part of the Woodward Corridor Initiative in Detroit, the Detroit Medical Center, Henry Ford Health System, and Wayne State University offer incentives for their employees to move to the Midtown neighborhood that surrounds the campuses and have established pilot programs to connect the institutions with local vendors and workforce training programs. Blue Cross Blue Shield of Michigan, Compuware, DTE Energy, Quicken Loans, and Strategic Staffing Solutions have also established financial incentives for their employees to rent or buy homes in or near downtown Detroit.[124]

- **Providing services and research on important local issues:** The University of Chicago's Urban Education Initiative operates four charter schools on Chicago's South Side; researches urban education issues; and trains teachers, social workers, and community school leaders.

- **Convening planning processes:** Kent State University in Ohio helped jumpstart plans for the Kent Gateway Multimodal Transportation Center by commissioning a feasibility study and supporting the city's application for a DOT grant, in part by contributing to the required local match.[125]

- **Contributing to transit and other infrastructure needs:** In Seattle, sponsors including Fred Hutchinson Cancer Research Center, University of Washington/UW Medicine, Evergreen Bank, Vulcan Real Estate, Pacific Place, Seattle Children's Hospital Research Institute, Pan Pacific Hotel Seattle, and Group Health provide up to 25 percent of the funds required to operate the South Lake Union streetcar line (Exhibit 32).

Exhibit 32. Seattle South Lake Union streetcar.
Source: Steve Morgan (Wikipedia).

- **Serving as anchor tenants for TOD projects:** In Atlanta, BellSouth helped make TOD at the Lindbergh City Center MARTA Station possible when, in the late 1990s, it consolidated several of its suburban offices into buildings at the station.

- **Orienting campus growth towards community goals:** San Jose State University built a new joint library with the city of San Jose to help overcome longstanding town-gown tensions and get the most out of limited city and university resources. Arizona State University is contributing to the revitalization of downtown Phoenix by building a new campus with the help of $232 million in municipal bonds.[126]

---

[124] Woodward Corridor Initiative, "Home page." http://www.woodwardcorridorinitiative.org. Accessed December 8, 2011.

[125] Kent City Manager Blog, http://www.kent360.com/category/city-university-stuff.

[126] Dittmar and Ohland (eds.), *The New Transit Town,* Island Press, 2004.

From the community's perspective, the potential benefits of involving anchor institutions in community and economic development are often clear: institutions can bring new funding sources to the table that are not directly tied to market conditions, tax revenue, or state or federal policy. They can also provide leadership capacity, research expertise, and other forms of human capital such as student or employee volunteers. Collaboration with anchor institutions can be particularly enticing in weak-market areas, where local government resources are strained. On the other hand, in some cases prior experiences could make a community reluctant to encourage anchor institution participation in local affairs. For example, tensions still linger in some communities from the 1960s and 1970s, when some large universities in declining inner-city neighborhoods conducted urban renewal programs that displaced residents or built physical barriers between their campuses and lower-income, minority neighborhoods.[127]

To effectively engage with anchor institutions, local governments must also understand the costs and benefits of community involvement from the institution's perspective. By investing in their neighborhoods, institutions can help create an environment that attracts employees, customers, and students; generate support from community and political leaders; contribute to an institutional mission; and potentially access new funding sources through municipal bonding capacity.[128] Institutions could be particularly interested in making TOD investments out of mission-related concerns for public health and the environment or if, for example, traffic congestion and parking capacity are hindering their expansion plans. However, designing and implementing a community investment strategy often requires significant time and attention from senior administrators and staff, not to mention the direct costs of contributing financially to infrastructure design or development or running a community program and the risks of failure and bad press.[129] To encourage institutions to invest in their communities, political and civic leaders need to recognize these costs and benefits and work to find areas of common interest.

GREATER UNIVERSITY CIRCLE INITIATIVE[130]

The Greater University Circle (GUC) Initiative illustrates how several institutions can come together with local government agencies to invest in critical local infrastructure needs and set the stage for new TOD.

University Circle is a 1 square-mile district about four miles east of downtown Cleveland. It is home to approximately 40 education, health, arts, and social services institutions (Exhibit 33). The district formed around Western Reserve University and the Case Institute of Technology (now consolidated into Case Western Reserve University), which located in University Circle in the 1880s.[131] Today, University Circle is also home to the Cleveland Museum of Art, the Cleveland Museum of Natural History, the Cleveland Botanical Gardens, University Hospitals (a major regional medical center), and dozens of other nonprofit organizations. The district is the fastest growing employment center in Cleveland, and the population of

---

[127] Webber, H. S. and Karlström, M. *Why Community Investment is Good for Nonprofit Anchor Institutions*. Chapin Hall at the University of Chicago. 2009.

[128] Ibid.

[129] Ibid.

[130] A version of this case study appears in *Fullerton Smart Growth 2030: FTC Specific Plan Funding & Financing Strategy & Case Studies* prepared for City of Fullerton and Southern California Association of Governments by Strategic Economics. 2012.

[131] University Circle, Inc. "History." http://www.universitycircle.org/about/history. Accessed December 8, 2011.

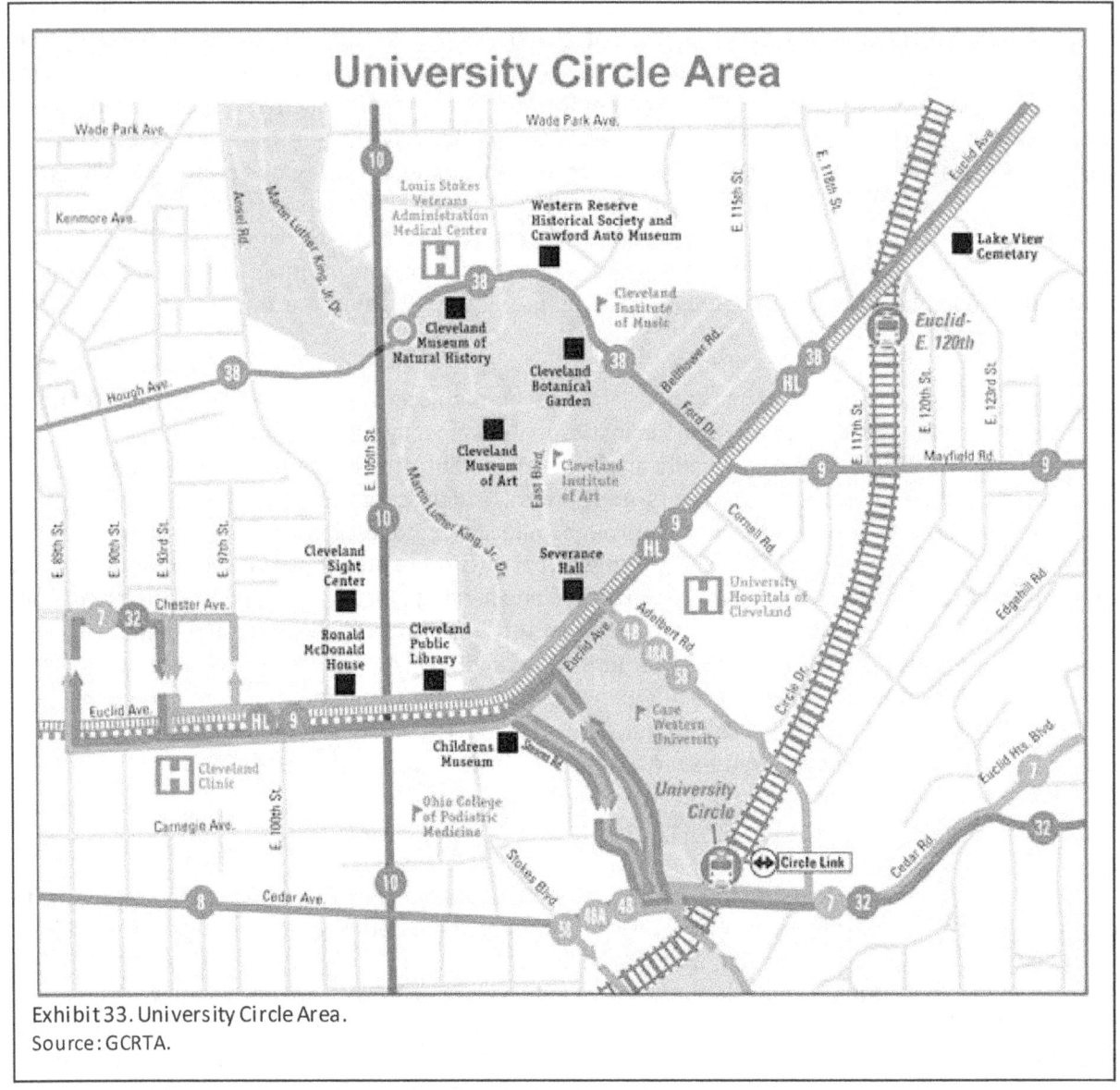

Exhibit 33. University Circle Area.
Source: GCRTA.

the University Circle census tract grew from 2,080 to nearly 3,680 between 2000 and 2010 while the city's total population shrank by 20 percent.[132] The surrounding communities, however, are some of the poorest and most disinvested in the metropolitan area.

University Circle is served by two stops on the Red Line, a decades-old rapid transit line operated by the Greater Cleveland Regional Transit Authority (GCRTA) that links the district with downtown Cleveland and the Cleveland Hopkins International Airport. In addition, a new bus rapid transit line known as the HealthLine provides a more direct route to downtown via Euclid Avenue, connecting to the Cleveland Clinic, performing arts centers, and other services and cultural institutions along the way.

---

[132] U.S. Census, 2000 and 2010. Census Tract 1187, Cuyahoga County.

Since the 1950s, many University Circle institutions have worked together to plan the district's development and provide services such as parking management, policing, and marketing through an organization called University Circle, Inc. (UCI) (see Exhibit 34). In 2005, the Cleveland Foundation initiated a new effort to bring the University Circle institutions together with institutions in the greater University Circle area, [133] GCRTA, and the city of Cleveland. The GUC Initiative aims to spur reinvestment in the struggling neighborhoods that surround the circle. Among other accomplishments, GUC has helped advance three major transportation projects and spurred significant residential, retail, and institutional development. The initiative has also created incentives to encourage employees of area nonprofits to live in the neighborhood and developed the Evergreen Cooperative Initiative, which encourages anchor institutions to purchase goods and services from local, worker-owned cooperatives. [134]

Exhibit 34. University Circle, Inc.

University Circle, Inc. (UCI), was founded in 1957 by civic leaders and philanthropists to administer the University Circle Master Plan, which laid out an orderly plan for institutional growth in the circle, and serve as a service organization to institutions in the district. Funded by an initial endowment of $7 million from a Cleveland philanthropist, the organization's original mission was to purchase and hold land for institutional expansion in the circle. UCI's purview quickly expanded to include providing districtwide services such as parking, shuttle buses, public safety, architectural review, and landscaping common areas. In the 1970s, UCI began working to strengthen the relationship between the circle's institutions and the surrounding neighborhoods by building housing and providing educational programs for local students.

Today, one of UCI's focus areas is developing land that the organization had originally purchased for institutional expansion to provide new housing, hotels, retail, and other services and amenities. UCI also plays an expanding role in infrastructure provision and maintenance. In 2005, UCI conducted a fundraising campaign that raised $7 million to make landscaping, wayfinding, and other improvements to Euclid Avenue, intended to augment streetscaping work that the city and GCRTA did as part of the bus rapid transit project. UCI also collects voluntary assessments from its member institutions to pay for streetscape cleaning and maintenance.

Sources: Personal communications with Chris Ronayne, Debra Berry, and Tom Mignogna, University Circle, Inc., December 8, 2011.

---

[133] Including parts of the city's Fairfax, Wade Park-Glenville, Hough, Little Italy, and Buckeye-Shaker neighborhoods, as well as Cleveland Clinic, the Veterans Affairs Medical Center, and several public schools.

[134] Cleveland was chosen as one of five sites for Living Cities' Integration Initiative (http://www.livingcities.org/integration/cities), which will provide up to $15 million in grants, loans, and program-related investments to expand the Evergreen Cooperative Initiative. For more information on the Evergreen Cooperatives, see: Howard, Ted, Lillian Kuri, and India Pierce Lee. "The Evergreen Cooperative Initiative of Cleveland, Ohio: Writing the Next Chapter for Anchor-Based Redevelopment Initiatives." Prepared for the Neighborhood Funders Group Annual Conference. September 29 - October 1, 2010. http://www.community-wealth.org/_pdfs/news/recent-articles/10-10/paper-howard-et-al.pdf.

TRANSPORTATION INFRASTRUCTURE PROJECTS

One of the first actions that the GUC Initiative took was to compile the growth plans of the various University Circle institutions and identify critical infrastructure needs. The initiative identified three transportation projects that were necessary to support the estimated $3 billion in development projects planned for the district. These projects would improve connections between University Circle and surrounding neighborhoods, encourage transit use, and spur TOD—goals that have become increasingly important to the circle's institutions as the district has become more congested and cost and lack of land have limited the potential for building new parking. The transportation projects are:[135]

- Reconfiguring a traffic circle, which had been a barrier to accessing University Circle from Rockefeller Park and neighborhoods to the north and west, to include bicycle and pedestrian paths.

- Relocating the Euclid-East 120th Red Line Station from an isolated area separated by freight rail tracks from University Circle to the heart of Little Italy and closer to Case Western Reserve University.

- Renovating the University Circle Rapid Transit Station to facilitate bus and rail transfers, improve pedestrian access, and bring the station into full compliance with accessibility standards.

To jumpstart the three projects, the Cleveland Foundation, UCI, and several of the larger institutions in the circle raised over $1 million to help GCRTA (which is responsible for the transit stations) and the Cuyahoga County Department of Public Works (which has jurisdiction over the traffic circle) plan the projects. This upfront funding for planning and design was critical to moving the projects forward. For example, GCRTA received $250,000 of the GUC Initiative's $1-million infrastructure fund to do a full planning study of the Euclid-East 120th Station, allowing the transit authority to hire consultants and determine how best to relocate the station to maximize opportunities for TOD.[136] GCRTA received a $12.5-million Transportation Investment Generating Economic Recovery grant in December 2011 to relocate the Euclid-East 120th Rapid Transit Station. GCRTA hopes to begin construction in 2013, as soon as it secures full funding for the $17.5-million project.[137]

Reconstruction of the University Circle Rapid Transit Station is expected to begin in 2012, funded by a $10.5-million Transportation Investment Generating Economic Recovery grant from DOT and a $2-million earmark. The Department of Public Works plans to begin construction on the traffic circle reconfiguration in 2013; construction will be funded with a combination of $3.275 million of federal safety and congestion program funds, $2.06 million from the Ohio Public Works Commission, $500,000 each from the city of Cleveland and the Cuyahoga County Engineer, and other sources.[138]

---

[135] Personal communication with Lillian Kuri, Program Director for Architecture, Urban Design, and Sustainable Development, Cleveland Foundation, by Alison Nemirow, Strategic Economics, December 8, 2011.

[136] Personal communication with Maribeth Feke, Director of Planning, Greater Cleveland Regional Transportation Authority, by Alison Nemirow, Strategic Economics, December 7, 2011.

[137] Greater Cleveland Regional Transit Authority. "Planning and Development." http://www.riderta.com/plandev. Accessed January 5, 2012.

[138] Cuyahoga County Department of Public Works. "Cuyahoga County Department of Public Works Proposes Traffic Circle and Roadway Configuration of Martin Luther King Jr. Drive and East 105th Street." News release. August 9,

TOD IN THE CIRCLE

University Circle is seeing a great deal of new development, largely on UCI-owned land. Much of the development is oriented around the Red Line stations or the HealthLine, which began operating in 2008. For example, a Cleveland developer is building a $44-million development that includes 102 apartments, 56,000 square feet of retail and restaurants, and a new Museum of Contemporary Art within a few blocks of the planned new site for the Euclid-East 120th Street Red Line Station.[139] UCI is also working with private developers on several market-rate apartment projects, a hotel, and an affordable housing project in neighboring East Cleveland.

Nearly all of the private development occurring in the circle—as in most of Cleveland—receives assistance from public agencies and/or foundations in the form of tax abatements, grants, low-cost loans, tax credits, and publicly provided infrastructure. However, the increasing interest from developers, national hotel operators, and other private entities in investing in the circle indicates that these investments in infrastructure and real estate development are building a market for TOD.[140]

LESSONS LEARNED

The GUC Initiative suggests several lessons for local governments interested in involving anchor institutions in developing TOD infrastructure or achieving other community and economic goals:

- **Anchor institution partnerships are long-term efforts that require extensive collaboration.** In University Circle, the institutions, supported by foundations and philanthropists, have taken the lead in organizing community investment strategies over many decades. In other places, mayors or other civic leaders might need to forge relationships with the leaders of anchor institutions to engage them in community and economic development efforts.

- **Anchor institutions might be able to fill critical gaps in infrastructure planning and development.** Educational, health, and arts institutions might have access to funding sources, such as endowments, grants, and alumni donations, that are not directly tied to market conditions, tax revenue, or state or federal policies. Thus, these institutions might be able to provide money for planning and design, better positioning projects to obtain public funding for construction. The GUC Initiative's grants to GCRTA, which enabled the transit agency to conduct more sophisticated planning studies and apply for Transportation Investment Generating Economic Recovery grants, show how an upfront investment can catalyze a project.

- **Institutions can provide the leadership required to convene political and community leaders and government agencies around common goals.** For example, relocating the Euclid-120th Street Station had been a longstanding goal for Case Western Reserve University, but the Little Italy community

---

2011; Northeast Ohio Areawide Coordinating Agency. "Improvements at East 105th Street/MLK Jr. Drive in Cleveland." 2008. http://www.noaca.org/105mlk.html. Accessed January 5, 2012.

[139] Schneider, K. "Cleveland turns Uptown into New Downtown." *New York Times*. November 29, 2011.

[140] Personal communication with Debra Berry, Vice President of Community Development, University Circle, Inc., and Tom Mignogna, Senior Director of Real Estate Development, University Circle, Inc., by Alison Nemirow, Strategic Economics, January 25, 2012.

had always resisted the idea. The GUC Initiative built consensus around the concept by providing funding for GCRTA to jointly plan the new station with the Little Italy Redevelopment Corporation.

- **Municipalities, transit authorities, and other government agencies can benefit not only from anchor institutions' direct investments in a neighborhood, but also from long-term results of those investments, such as new jobs and residents, an enhanced tax base, and increased transit ridership.** While most nonprofit institutions do not pay property taxes, their investments in local infrastructure, catalytic development projects, community services, and economic revitalization can have long-term benefits for the public sector. Although development in University Circle still largely requires public support, the ongoing public and nonprofit investments in infrastructure and property development are gradually building a market for private development.

APPLICABILITY TO OTHER PLACES

This model works best in areas with institutions that have strong local ties and a long-term interest in their communities' wellbeing. For these institutions, the benefits of infrastructure or other investments must outweigh the costs of staff time, political capital, and dollars. Institutions and communities could reap the greatest benefits from institutional investment in weak-market places where local government resources are scarce, other options for financing infrastructure (e.g., value capture strategies) are difficult to apply, and infrastructure needs are particularly pressing and likely to affect an institution's ability to attract employees, customers, or students. However, the model would also be effective in strong-market areas.

ENTITIES INVOLVED IN IMPLEMENTING THE MODEL

Unlike more traditional infrastructure financing tools, public agencies cannot simply develop a project proposal on their own and apply to an institution for funding. Instead, developing an anchor institution partnership is a long-term effort that requires civic, community, and institutional leaders to work together to identify areas of common interest and reach compromises where institutional and community goals conflict.

- Mayors and other local government officials can:

  o Incorporate and engage institutions in local economic development strategies.

  o Convene institutional leaders regularly with business, foundation, and other civic leaders to identify partnership opportunities.

  o Establish a liaison office to build relationships with institutions.

- Foundations, community leaders, and business leaders can:

  o Involve institutions in community and business forums and public-private initiatives.

  o Seek partnerships with institutions that benefit all parties (e.g., in real estate development and workforce training).

- Institutional leaders can:

    o Create strategies based on meaningful community participation for investing in the surrounding communities.

    o Assign specific departments to implement economic development goals, and create high-level positions to coordinate community engagement and other economic development efforts.

    o Ask board members and senior administrators to serve on boards of local business associations and philanthropic organizations.[141]

REFERENCES

CEOs for Cities. *How to Behave Like an Anchor Institution: A White Paper by CEOs for Cities with Living Cities.* June 2010. http://www.ceosforcities.org/pagefiles/behave_anchors.pdf.

CEOs for Cities. *Leveraging Anchor Institutions for Urban Success.* August 2007. http://www.ceosforcities.org/pagefiles/CEOs_LeveragingAnchorInstitutionsforUrbanSuccess_FINAL.pdf

Cleveland Foundation. "Greater University Circle: Cleveland's Urban Core." http://www.clevelandfoundation.org/VitalIssues/NeighborhoodsAndHousing/GreaterUniversityCircle. Accessed December 8, 2011.

Cuyahoga County Department of Public Works. "Cuyahoga County Department of Public Works Proposes Traffic Circle and Roadway Configuration of Martin Luther King Jr. Drive and East 105th Street." News release. August 9, 2011. http://publicworks.cuyahogacounty.us/pdf_publicworks/en-US/TrafficCircleNews.pdf

Dittmar, Hank and Ohland, Gloria (eds.) *The New Transit Town,* Island Press, 2004.

Greater Cleveland Regional Transit Authority. "Planning and Development." http://www.riderta.com/plandev. Accessed January 5, 2012.

Initiative for a Competitive Inner City and CEOs for Cities. *Leveraging Colleges and Universities for Urban Economic Revitalization: An Action Agenda.* 2002. www.ceosforcities.org/files/colleges_1.pdf.

Jarboe McFee, Michelle. "Developer MRN Ltd. Secures Financing for First Phase of Uptown Neighborhood Project at University Circle." *Cleveland Plain Dealer.* April 8, 2010. http://www.cleveland.com/business/index.ssf/2010/04/mrn_ltd_secures_financing_for_fir.html.

Northeast Ohio Areawide Coordinating Agency. "Improvements at East 105th Street/MLK Jr. Drive in Cleveland." http://www.noaca.org/105mlk.html. Accessed January 5, 2012.

Personal communication with Maribeth Feke, Director of Planning, Greater Cleveland Regional Transportation Authority, by Alison Nemirow, Strategic Economics, December 7, 2011.

Personal communication with Lillian Kuri, Program Director for Architecture, Urban Design, and Sustainable Development, Cleveland Foundation, by Alison Nemirow, Strategic Economics, December 8, 2011.

---

[141] Initiative for a Competitive Inner City and CEOs for Cities. *Leveraging Colleges and Universities for Urban Economic Revitalization: An Action Agenda.* 2002.

Personal communication with Chris Ronayne, President, University Circle, Inc., by Alison Nemirow, Strategic Economics, December 8, 2011.

Personal communication with Debra Berry, Vice President of Community Development, University Circle, Inc., and Tom Mignogna, Senior Director of Real Estate Development, University Circle, Inc., by Alison Nemirow, Strategic Economics, January 25, 2012.

Schneider, Keith. "Cleveland turns Uptown into New Downtown." *New York Times*. November 29, 2011. http://www.nytimes.com/2011/11/30/realestate/commercial/cleveland-ignites-job-growth-with-rebuilding-project.html.

University Circle, Inc. "History." http://www.universitycircle.org/about/history. Accessed December 8, 2011.

Webber, Henry S. and Mikael Karlström. *Why Community Investment is Good for Nonprofit Anchor Institutions.* Chapin Hall at the University of Chicago. 2009. http://www.chapinhall.org/research/report/why-community-investment-good-nonprofit-anchor-institutions.

Woodward Corridor Initiative, "Home page." http://www.woodwardcorridorinitiative.org. Accessed December 8, 2011.

B. CORRIDOR-LEVEL PARKING MANAGEMENT

A corridor-level parking management model would set parking prices and manage parking demand across a transit corridor or system, including both transit station parking and surrounding on- and off-street spaces. Revenue from parking fees throughout the system could be pooled to finance structured parking or other improvements at strategic locations, generating more revenue than a station-by-station approach and reducing the incentive for commuters to drive to a station or a neighborhood street where they can park for free. No region appears to have implemented this type of comprehensive strategy to date. However, in 2010 the city of Aurora in the Denver metropolitan area commissioned a study that lays out an innovative parking management strategy for the planned I-225 light-rail corridor, including a proposal for pricing on- and off-street parking throughout the corridor and using the proceeds to finance structured parking at targeted light-rail stations.

To help understand how a corridor-level parking management strategy could be used to fund parking for TOD, this section discusses:

- The role that corridor-level parking management could play in facilitating TOD.

- The Aurora plan, which illustrates the components required to make a corridor-level parking plan work, including comprehensive demand management, parking fees, other potential revenue sources, and steps for implementation.

- Lessons learned from the Aurora plan and other considerations for cities and transit agencies to consider in implementing corridor-level parking management.

THE ROLE OF CORRIDOR-LEVEL PARKING MANAGEMENT IN FUNDING AND FINANCING TOD INFRASTRUCTURE

Financing structured parking is one of the most difficult challenges associated with TOD. However, in many transit station areas, successful TOD requires structured parking. Providing parking in structured garages instead of surface lots can increase parking capacity while leaving more land for housing and commercial space. Many transit agencies own large surface parking lots that could be prime sites for TOD. Converting these lots to TOD would help achieve regional housing and smart growth goals and could boost transit use by putting more people within walking distance of transit.[142] However, building TOD on these lots would remove commuter parking spaces, and few transit agencies have the resources to build structured parking on their own, initial funding for new transit systems rarely includes money to build structured parking, and few state or federal transportation funding sources can be used to pay for structured parking. And, except for in the hottest real estate markets, development projects rarely generate sufficient value to cover the cost of building both replacement parking for the station and the parking to serve the development.

---

[142] Nelson|Nygaard. *Transit Agency Parking Pricing and Management Practices: Peer Review.* Prepared for the Denver Regional Council of Governments. 2010.

Another complication is that parking at many transit stations, as well as on surrounding streets and in nearby parking facilities, is either free or priced below the cost of building and operating the parking). Free or inexpensive parking encourages travelers to drive alone to transit stations instead of walking, bicycling, taking a bus or shuttle, or carpooling.[143] Charging for parking, could both reduce demand for parking by encouraging travelers to take alternative modes and create a revenue source to fund structured parking or other improvements, especially if combined with improving bicycle, pedestrian, and bus access to the station and increasing the supply of housing, retail, and offices in the surrounding area. From a transit agency's perspective, charging for parking could also help promote off-peak use.

Since free parking is first-come, first-served, commuters want to arrive before the last parking space is taken. As a result, many park-and-ride stations experience a sharp peak in parking demand during the morning commute. During the rest of the day, the parking lot remains full of commuters' cars—and the station platform sits empty because no one else can find a parking space. Pricing parking, ideally at variable rates depending on the time of day, can ensure that some parking spaces are always available.

Several cities have begun implementing comprehensive parking strategies that manage demand and raise revenue for local transportation improvements (see Exhibit 35), while some transit agencies have also begun to experiment with pricing parking (see Exhibit 36).

THE AURORA STRATEGIC PARKING PLAN AND PROGRAM STUDY[144]

The city of Aurora is an eastern suburb of Denver and the second largest city in the Denver region. Although Aurora's population (325,000 in the 2010 Census) is just over half the size of Denver's, Aurora encompasses approximately the same land area as the larger city and includes parts of three counties. Much of the developed land in Aurora is occupied by relatively low-density housing, and the city has many tracts of vacant and underused land.

Aurora currently has one light-rail station, Nine Mile Station, which opened in 2006 and is the terminus of the Southeast Corridor Line. The Regional Transportation District (RTD) of Denver has a plan called FasTracks that will greatly expand the city's transit network. The planned I-225 Corridor Line will connect Nine Mile Station to the planned East Corridor light-rail line and to I-70, one of the region's primary highways. All eight new stations on the I-225 Corridor Line, as well as at least one additional station (40th/Airport) on the East Corridor Line, will be in Aurora. The East Corridor Line is projected to open in 2016; the completion date of the I-225 Corridor Line is uncertain.[145] The existing land uses around the

---

[143] Shoup, D. *The High Cost of Free Parking*. American Planning Association. 2005.

[144] Unless otherwise stated, the source for all of the information in this section is: Wilbur Smith Associates in association with URS Corporation, Michael R. Kodama Planning Consultants, and Rick Williams Consulting. *City of Aurora Strategic Parking Plan and Program Study*. Prepared for the city of Aurora. 2010. A version of this case study appears in *Fullerton Smart Growth 2030: FTC Specific Plan Funding & Financing Strategy & Case Studies* prepared for City of Fullerton and Southern California Association of Governments by Strategic Economics. 2012.

[145] Although final design and environmental review has been completed for the I-225 Corridor Line, the RTD board has postponed construction due to a shortfall in revenue from the sales tax that funds the FasTracks program. The 2020 completion date for the I-225 Corridor Line was contingent upon voters approving a second sales tax increase in 2012 November to cover the shortfall. However, RTD decided not to seek voter approval for the sales tax increase in 2012. When the Aurora parking plan was completed in June 2010, the I-225 Corridor Line was expected to open in 2015.

planned stations vary, but they include a lot of underused and vacant land, as well as a mix of low-density residential, commercial, and industrial space.

Exhibit 35. Tools for City Parking Management

Cities like Pasadena, California, and Portland, Oregon, are becoming known for their comprehensive, district-based approaches to parking management. These programs combine tools to use existing commercial parking spaces efficiently; protect residential neighborhoods from visitor, employee, and/or commuter parking; and raise funds for transportation and other neighborhood improvements. The key benefit of these district-wide programs is that, by dedicating revenue from parking fees to local improvements, cities can convince property and business owners to accept parking meters and other pricing.

Common types of parking fees include:

- **User fees:** Charging drivers for on- and off-street parking.

- **In lieu fees:** Charging developers a one-time fee as a condition for opting out of a portion of the minimum parking requirement.

- **Transportation or parking impact fees:** Charging developers a one-time fee proportional to the development's impact on the transportation system or parking supply.

- **Parking tax or assessment:** Charging parking lot owners an annual fee for each stall or charging sales tax on parking fees. (These are typically used in downtowns, where commercial lots that charge for parking are more common.)

Fees are often combined with other management strategies, such as:

- **Residential permit parking:** Restricting long-term parking in residential neighborhoods to that neighborhood's residents.

- **Time limits:** Limiting how long people can use on-street parking (e.g., five-minute loading zones, two-hour parking zones).

- **Wayfinding signage and real-time parking information**: Providing information to help drivers understand parking availability and location.

- **Shared parking:** Allowing different uses to share parking facilities.

- **Transportation demand management strategies:** Free or reduced-cost transit passes, bicycle and pedestrian improvements, and shuttles.

Sources: Tumlin, Jeffrey. *Sustainable Transportation Planning*. John Wiley & Sons, Inc. 2012; Personal communication with Rick Williams, Principal, Rick Williams Consulting, by Alison Nemirow, Strategic Economics, December 8, 2011; Shoup, Donald. *The High Cost of Free Parking*. American Planning Association. 2005.

Exhibit 36. Transit Agencies and Parking Fees

Transit agencies across the country are experimenting with pricing parking. Most of the agencies that currently charge for parking are well-established, heavy-rail systems that serve large cities such as San Francisco; Washington, D.C.; Atlanta; Chicago; and Los Angeles. However, smaller systems such as TriMet in Portland, Oregon; the Regional Transit District in Sacramento, California; and RTD in Denver are beginning to implement parking fees.

As agencies have begun to charge for parking, they have encountered some similar challenges, all of which would affect a region's ability to implement corridorwide pricing:

Paradigm shift: Many customers have come to expect free parking. Understanding that providing parking involves costs can help change this mindset. Instead of assuming everyone will reach transit stations by driving alone, transit agencies can encourage walking, biking, carpooling, and using buses.

Political opposition: Board members with suburban constituents who are often used to driving alone and parking for free are particularly likely to oppose pricing.

Statutory or regulatory barriers: RTD has strict statutory limitations on its ability to manage and price parking. FTA regulations can also affect parking management. Transit agencies argue that the New Starts ridership and cost-effectiveness models effectively reward providing free parking at the expense of alternative modes of access and that FTA's guidelines on how much replacement parking is required in joint development projects that involve land purchased with federal money are confusing and impede TOD.

Prioritizing expenditures: Most parking fees flow into agencies' general funds, although some agencies have considered dedicating a portion of parking revenue to improving alternate modes of access at the stations where the revenue is generated. Transit riders might be more likely to support paying for parking if they can see tangible benefits at their station.

Sources: Nelson|Nygaard *Transit Agency Parking Pricing and Management Practices: Peer Review.* Prepared for the Denver Regional Council of Governments. 2010; U.S. Government Accountability Office. *Public Transportation: Federal Role in Value Capture Strategies for Transit is Limited, but Additional Guidance Could Help Clarify Policies.* 2010.

The I-225 Corridor Line is expected to draw park-and-ride commuters from across the southeast metropolitan area. The FasTracks program is budgeted to build surface parking lots on the I-225 Corridor Line. Concerned that parking demand would significantly exceed the supply that FasTracks will provide, the city of Aurora commissioned the *Strategic Parking Plan and Program Study* to estimate parking demand and identify corridorwide parking management strategies that would "support the city's land use vision and station area plans, maximize efficient use of parking spaces, preserve and enhance the economic vitality and quality of life, and protect surrounding neighborhoods and businesses from spillover commuter parking."[146] The city asked the consultant team to consider strategies for financing structured parking garages in certain locations because RTD does not have the budget to build additional parking. RTD has been open to working with cities to explore creative options for developing and managing station parking (subject to state-imposed restrictions discussed later).

---

[146] Wilbur Smith Associates in association with URS Corporation, Michael R. Kodama Planning Consultants, and Rick Williams Consulting. *City of Aurora Strategic Parking Plan and Program Study.* Prepared for the city of Aurora. 2010. p ES-3.

Based on the Denver Regional Council of Government's travel demand model and research on comparable transit systems around the country, the plan estimated that, absent pricing or supply constraints, demand for parking on the I-225 Corridor Line could be between 3,300 and 4,400 spaces on opening day and could be as high as 9,000 spaces by 2035. Building sufficient parking to absorb all of this demand would be cost-prohibitive, so the plan recommended implementing a set of parking demand management tools, including pricing, and building structured parking only in four locations where commuter demand is expected to be highest.

DEMAND MANAGEMENT

The plan recommended that the city establish mechanisms to manage parking and enforce parking regulations in all the I-225 Corridor Line station areas. A comprehensive demand management strategy that encompasses both on-street spaces and off-street parking lots makes pricing parking at the station areas more feasible and ensures that spaces on neighborhood streets are available for residents, shoppers, visitors, and employees.

To manage on-street parking, the city would first implement time-limited parking in commercial districts, coupled with a residential parking permit program that allows residents to park longer. To help manage off-street parking spaces efficiently (and to set the stage for some of the revenue-generating mechanisms discussed below), the plan recommended that the city begin to centralize management of new commercial parking spaces that are built in the corridor, perhaps using an integrated database or even bringing them under direct city management through development agreements. The plan also recommended that the city encourage shared parking arrangements; consider installing wayfinding and real-time parking information systems to direct drivers to open parking spaces; and improve pedestrian, bike, and bus access to the station to reduce parking demand.

In commercial areas where on-street parking use remains above 85 percent after the demand management measures described above are implemented, the plan recommended that the city implement a pricing system (i.e., meters) that prioritizes on-street parking immediately in front of stores for shoppers and encourages employees and other long-term visitors to park in off-street facilities or on streets farther from the core retail areas where parking is cheaper.

PRICING STATION PARKING

Charging for parking at stations is a critical component of the plan, but one that would be difficult to implement. Charging for park-and-ride parking would generate funds to build structured parking and help manage demand by encouraging transit riders to get to the station by walking, bicycling, taking a bus or shuttle, or carpooling instead of driving. However, Colorado law restricts RTD from charging for parking, with the narrow exceptions of vehicles registered to owners outside of the RTD taxing district,[147] vehicles parked more than 24 hours, and transit users who pay in advance to reserve parking

---

[147] State law gives RTD the authority to access vehicle registration information to determine whether park-and-ride users live outside the taxing district.

spaces during peak hours. The plan noted that this policy "would make management, revenue collection, and enforcement of parking charges more complex."[148]

The plan estimated that if the state restrictions were lifted, RTD could charge $1.25 per parking space per day without reducing ridership.[149] At this price, the agency could raise about $500 per space per year, up from about $340 per space per year under the current pricing policy. This additional revenue would help cover the cost of building and operating structured parking, although a significant amount of additional revenue would still be required.

Exhibit 37 shows the pro forma analysis that the consultant team conducted, which estimated the cost of building the desired structured parking at $20,470 per stall, including land acquisition and construction, plus $236 per stall per year in operating costs. The total annual cost over 10 years of building the structures comes to more than $3.58 million. (By comparison, providing the same amount of parking in surface lots would cost about $9,260 per stall, with operating costs also slightly lower at $200 per stall per year, for a total annual cost of about $1.82 million). Under RTD's current policies, the 2,145 parking spaces in the four structures would generate $728,000 per year, leaving a gap—the difference between annual costs and revenue from parking fees—of $2.85 million. Charging all users (at $1.25 per day) could increase parking revenue to $1.085 million per year, reducing the gap marginally to $2.5 million. Although the plan did not consider this alternative, raising the daily price to $4.15 per user could generate $3.6 million and fully close the gap, assuming that demand for parking did not decline as a result of the higher price.

	Structured Parking	Surface Parking
Total development cost (land plus construction)	($43,908,867)	($19,862,680)
Average development cost per stall	($20,470)	($9,260)
Annual debt service	($3,080,249)	($1,393,383)
Annual operating costs	($505,946)	($8,383)
Average annual operating cost per stall	($236)	($200)
Total annual cost (debt service plus operating cost)	($3,586,195)	($1,821,766)
Annual gross revenue from parking charges		
With current RTD rates	$728,475	$728,475
With charges assessed to all users	$1,085,684	$1,085,684
Remaining annual gap (costs minus revenues)		
With current RTD rates	($2,857,720)	($1,093,291)
With charges assessed to all users	($2,500,511)	($736,082)

Exhibit 37. Summary of pro forma analysis: costs, revenue, and remaining "gap" for 2,145 structured versus surface parking stalls at four sites.
Source: Wilbur Smith Associates in association with URS Corporation, Michael R. Kodama Planning Consultants, and Rick Williams Consulting. *City of Aurora Strategic Parking Plan and Program Study.* Prepared for the city of Aurora. 2010. Table 5-6.

---

[148] Wilbur Smith Associates in association with URS Corporation, Michael R. Kodama Planning Consultants, and Rick Williams Consulting. *City of Aurora Strategic Parking Plan and Program Study.* Prepared for the city of Aurora. 2010. p V-3.

[149] At this price, a commuter's combined out-of-pocket cost for a monthly transit pass and park-and-ride parking would be no more than 60 percent of the total cost of driving to and parking in downtown.

RAISING ADDITIONAL REVENUE TO FUND STRUCTURED PARKING

Given the gap between the annual costs of financing structured parking and the revenue that parking fees could generate, other funding sources will be required to pay for the structured parking. The plan recommended that the city consider implementing one or more of the following mechanisms throughout the entire I-225 corridor:

- A parking fee in lieu option, which would allow developers to opt out of a portion of minimum parking requirements in exchange for paying a fee.

- A one-time transportation impact fee on new development.

- An annual flat fee (in the range of $5 to $15 per space) on all commercial parking stalls in the corridor.

All three of these fees would spread the cost of transportation improvements over many users, creating a significant source of revenue without being burdensome to individuals or developers. On the other hand, the plan acknowledged that the revenue generated by these fees—or by other mechanisms such as special assessment districts—could probably not be spent on a traditional park-and-ride facility, which primarily benefits commuters who neither live nor work near the station. Any parking facility funded with these fees would have to be available to the public to justify spreading the cost to developers and property owners.[150] However, managing a shared parking facility that provides parking for other uses as well as transit riders can be challenging if the peak parking occupancy hours for transit riders (approximately 7 or 8 a.m. to 6 p.m.) overlap with the peak parking hours for the use that is sharing the parking, such as offices (9 a.m. to 5 p.m.) or stores (10 a.m. to 2 p.m.).[151] The city of Aurora would need to work closely with RTD to identify locations where a shared parking arrangement might work.

The plan did not include an analysis of how much the three potential fees might raise, but it assumed that, given the limitations on how revenue could be spent and the difficulties of managing shared parking, the city would need to ask voters to extend the life of an existing tax levy and issue new debt to build structured parking. The plan estimated that the city could raise $7.5 to $8.13 million for parking development by issuing new bonds.

IMPLEMENTATION

The plan recommended that the city begin by clearly defining its role in financing, building, and managing parking. That role should include city involvement in initiating and implementing new funding mechanisms, acquisition of land, negotiations with potential funding partners, operations and management of both on- and off-street parking, and enforcement of parking regulations. After defining the appropriate city role in providing parking, the plan recommended that in the near term (the next 48 months), the city focus on:

---

[150] To establish a transportation impact fee, the city would be legally required to establish a direct nexus between the development paying the fee and the benefit that the fee provides.

[151] Personal communication with Rick Williams, Principal, Rick Williams Consulting, by Alison Nemirow, Strategic Economics, December 8, 2011.

- Working with RTD and the state legislature to revise limits on charging for parking at park-and ride facilities.

- Initiating on-street parking management in station areas.

- Asking voters to issue new debt for parking facilities.

- Establishing the parking fee in lieu option.

In the mid-term (26 to 60 months), the plan recommended that the city explore instituting either the transportation impact fee on new development or the flat fee on all commercial parking spaces.

To fully implement the plan's recommendations over the long term, the city would need to work with RTD and local stakeholders to create operating agreements and financing plans for each station. For example, at stations where local property and business owners are particularly interested in parking management, the plan envisioned that the city, RTD (assuming the agency had the authority under state law), and local stakeholders could form a parking district authority to jointly manage parking or a business improvement district to fund a shared parking facility. At other stations where the city and RTD would both contribute funding for park-and-ride facilities, they could sign an agreement governing which agency would operate the facility and how surplus revenue would be allocated.

As of early 2012, the Aurora City Council supported the idea that the city should play a role in managing parking, and RTD was working to bring a bill before the state legislature that would allow the agency to charge more riders for parking through a third-party entity. The city was also working on including funding for structured parking in a November ballot measure that will ask voters to decide whether the city should issue new bonds for infrastructure improvements and beginning to reach out to business owners in the station areas.[152]

LESSONS LEARNED

The potential for a city or region to manage parking at the corridor-level is not tied directly to the strength of the local real estate market, but rather to localities' and transit agencies' capacity and political will for working together to impose parking prices and coordinate parking policy across jurisdictional boundaries. Corridor-level parking management presents an opportunity to manage demand for parking comprehensively across station areas and generate parking fees and other revenue from a broad base. The parking plan developed for Aurora illustrates some issues that other cities and transit agencies would need to consider to implement a corridor-level parking management strategy:

- **A comprehensive approach is required to create a market for priced commuter parking and create opportunities for cross-subsidy among different types of parking.** Charging for or restricting access to surrounding on- and off-street parking is critical to charging for parking at transit stations because it removes free on-street options for commuters and keeps those spaces free for neighborhood

---

[152] Personal communication with John Fernandez, Manager of Comprehensive Planning, City of Aurora, and Huiliang Liu, Principal Transportation Planner, City of Aurora, by Alison Nemirow, Strategic Economics, December 2011 and February 2012.

residents. A corridor-level approach to pricing could also help ensure that parking policies are consistent for all stations on the line.

- **Close coordination among multiple entities is required to make corridor-level parking management work.** Aurora has an advantage over most jurisdictions in that the I-225 Corridor Line falls entirely within its borders, allowing the city not only to set comprehensive pricing and management policies, but also to think strategically about the best places for structured parking. For a corridor-level approach to work in most places, multiple jurisdictions would have to work closely with each other and with the transit authority to identify the best locations for structured parking, spread the cost across a wide base, and ensure that parking is priced consistently across jurisdictions so that all users pay their fair share.

- **A corridor-level parking management plan could allow revenue-sharing throughout the corridor.** The plan emphasized the advantages of spreading the cost of building parking over as wide a base as possible through some combination of user fees, a parking tax or assessment on all commercial parking stalls in the corridor, in lieu or impact fees on all new development in the corridor, and debt financed by a citywide tax levy. In addition, although the plan developed for Aurora does not fully explore this possibility, corridor-level parking management could allow cities and transit agencies to pool user fees from multiple parking facilities to pay for structured parking or other improvements in targeted locations.

- **Statutory, regulatory, and other barriers could restrict transit agencies' ability to charge for parking.** Exhibit 36 discusses some of the challenges that transit agencies have faced as they begin to charge for parking. Transit agencies need not only the statutory authority to set prices, but also the political will. RTD has asked the state legislature to lift restrictions on the agency's ability to charge for parking. For the Aurora plan to work, RTD would need not only to obtain the necessary statutory authority to enact widespread parking charges, but also to agree that revenue from parking fees should be dedicated to financing structured parking rather than funding other station-access improvements or going into the agency's general fund. It is not clear if RTD would be amenable to dedicating funds in this way. The agency's parking demand estimates for the I-225 Corridor Line are lower than Aurora's, and RTD prioritizes pedestrian, bus, and bicycle access to stations over vehicle parking. [153]

- **Determining the "right" price for parking depends on whether the goal is to ensure that consumer demand roughly matches supply (the "market" price) or to cover the cost of building and operating parking (the "production" price).** Finding the appropriate market price requires flexibility and trial and error, and the market price might not always be high enough to cover production costs. To set a true market price at each station—the price where consumer demand roughly matches supply and a few parking spaces (roughly 15 percent) are available at all times—cities and transit agencies typically experiment with moving prices up and down. [154] If a parking lot is open to the general public in addition to commuters, the market price is likely to change as development

---

[153] RTD. *RTD Transit Access Guidelines.* 2009.

[154] Nelson|Nygaard. *Transit Agency Parking Pricing and Management Practices: Peer Review.* Prepared for the Denver Regional Council of Governments. 2010.

occurs around the stations.[155] In places like Aurora, where parking has traditionally been free or very cheap, the market price could (at least initially) be lower than the production price, and other funding and financing sources would be required to cover the cost of building parking. By setting the price of parking below the production price to ensure that parking facilities are roughly 85 percent full at all times, a city or transit agency in effect subsidizes parking, presumably to meet other goals like encouraging transit use.[156]

APPLICABILITY TO OTHER PLACES

The potential for localities and transit agencies to manage parking at the corridor level is not tied directly to the strength of the local real estate market, but rather to their capacity and political will for working together to coordinate parking policy across jurisdictional boundaries and set prices on parking at transit stations, on streets, and in other parking facilities. Depending on a region's needs and the amount of funding raised, a corridor-level approach to parking management could raise revenue to finance not just structured parking, but also (or alternatively) other transportation needs, or even non-transportation infrastructure needs like stormwater management.

ENTITIES INVOLVED IN IMPLEMENTING THE MODEL

Few places have the ability to set prices and manage parking across a corridor in (relative) isolation. In general, municipalities and transit agencies would need to work together closely to establish a corridor-level parking management plan. Even without a corridor-level approach, some of the policy changes described below could better manage parking demand, reduce congestion, encourage off-peak transit use, and enable TOD.

- After creating a plan, **localities and transit agencies** could either implement it through their respective public works or parking management departments or form a joint authority to manage station parking facilities and other on- and off-street spaces.

- Depending on the scope of the plan, **business and property owners** could also play a role in implementation.

- **Transit agencies** could consider:

   o Prioritizing other modes of access over park-and-ride, including walking, bicycling, taking buses or shuttles, and carpooling.

   o Reducing requirements for one-to-one replacement parking.

   o Sharing parking with other uses.

   o Charging for parking.

---

[155] Personal communication with Rick Williams, Principal, Rick Williams Consulting, by Alison Nemirow, Strategic Economics, December 8, 2011.

[156] Tumlin, J. *Sustainable Transportation Planning*. John Wiley & Sons, Inc., 2012. p. 203.

- **Cities** could begin to implement transportation and parking demand management strategies around station areas and in other places with high parking demand through mechanisms such as:

  o Time limits.

  o Wayfinding and real-time parking information.

  o Bicycle, pedestrian, and other access improvements.

  o Residential permit parking districts.

  o On- and off-street parking pricing.

- **State and regional governments** could:

  o Lift restrictions on transit authorities and other agencies' ability to manage parking and set prices.

  o Provide technical assistance for jurisdictions considering parking management strategies.

  o Set regionwide parking standards (in some cases).[157]

REFERENCES

Nelson|Nygaard. *Transit Agency Parking Pricing and Management Practices: Peer Review*. Prepared for the Denver Regional Council of Governments. 2010. http://tod.drcog.org/resources/resourcesdatabase/transit-agency-parking-pricing-and-management-practices-peer-review.

Personal communication with Jeffrey Boothe, Partner, Holland and Knight, by Alison Nemirow, Strategic Economics, December 15, 2011.

Personal communications with John Fernandez, Manager of Comprehensive Planning, City of Aurora, and Huiliang Liu, Principal Transportation Planner, by Alison Nemirow, Strategic Economics, December 2011 and February 2012.

Personal communication with Bill Sirois, Manager of Transit-Oriented Development, RTD, by Alison Nemirow, Strategic Economics, December 15, 2011.

Personal communication with Jeffrey Tumlin, Owner, Nelson|Nygaard, by Alison Nemirow, Strategic Economics, December 5, 2011.

Personal communication with Rick Williams, Principal, Rick Williams Consulting, by Alison Nemirow, Strategic Economics, December 8, 2011.

RTD. *RTD Transit Access Guidelines*. 2009. http://www3.rtd-denver.com/content/DesignCriteria/TransitAccessGuidelines/Transit_Access_Guidelines_Final.pdf.

Shoup, Donald. *The High Cost of Free Parking*. American Planning Association. 2005.

Tumlin, Jeffrey. *Sustainable Transportation Planning*. John Wiley & Sons, Inc. 2012.

---

[157] For example, Metro, the Portland region's MPO, requires local governments to adopt lower parking requirements in areas with good transit service.

U.S. Government Accountability Office. *Public Transportation: Federal Role in Value Capture Strategies for Transit is Limited, but Additional Guidance Could Help Clarify Policies.* 2010. http://www.gao.gov/products/GAO-10-781.

Wilbur Smith Associates in association with URS Corporation, Michael R. Kodama Planning Consultants, and Rick Williams Consulting. *City of Aurora Strategic Parking Plan and Program Study.* Prepared for the city of Aurora. 2010. https://www.auroragov.org/stellent/groups/public/documents/article-publication/448199.pdf.

C. LAND BANKING FOR TOD INFRASTRUCTURE

Land assembly and acquisition can be major challenges for TOD because often there is little developable land near transit and what land is there generally costs more. Some types of TOD require relatively large parcels, but infill locations often have smaller parcels with scattered ownership. During the real estate market boom of the 2000s, real estate speculation as transit lines were planned and constructed pushed land prices higher. In a strong real estate market, developers face competition for good development sites near transit. In weaker markets, TOD developers can find it even more difficult to gain control of property because of the difficulty of getting short-term financing and uncertainty about when long-term project financing will be secured. Regardless of market strength, the challenges of land acquisition can be compounded for TOD that includes affordable housing because it can be more difficult to get financing to secure sites for affordable housing due to underwriting criteria.

Because transit stations are planned and built over many years, land and property values can begin to rise even before the new station opens. To keep projects financially feasible and enable construction of affordable housing, local governments are seeking tools to acquire and assemble land before it becomes too expensive. In response to the widespread challenge of land acquisition in transit station areas, various tools and strategies are emerging across the country. Land banking, which has used for decades, could be one solution to the challenges of land assembly and acquisition for TOD.

TRADITIONAL MODEL OF LAND BANKING

Traditional land banks are typically public authorities created to acquire vacant, underused, or tax-foreclosed properties from government agencies, nonprofits, or private owners. Land banks acquire distressed properties to stabilize neighborhoods or create affordable housing. They are usually located in weaker real estate markets where market demand is not adequate for the properties to be redeveloped without government intervention and where vacant and abandoned properties are widespread and causing problems that need to be remedied.[158]

Although traditional land banking strategies have been used extensively throughout the country, few places have tried to connect land banking with facilitating new development around existing or planned transit stations. The traditional land banking model is unlikely to be feasible in a real estate market with high land values or growing demand because of the high costs of holding land that could be developed without government intervention.

MODELS OF LAND BANKING FOR TOD

To use land banking to encourage TOD, a land bank could acquire land in an area that is not yet ready for development, either because it is a weaker real estate market, or because the transit has not yet arrived. The land bank would then hold (or "bank") the land until appropriate development is possible. In this scenario, land bank authorities would acquire properties before land prices increase and save the property to be developed as affordable or mixed-income housing when transit service begins or when market demand increases. In many cases, a land banking authority would "write down" the value of the land when transferring it to a developer, meaning that the land bank sells the land to the developer at a

---

[158] Alexander, F. *Land Banks and Land Banking.* Center for Community Progress. June 2011.

reduced price or transfers the land at no cost. This method helps keep projects financially feasible by allowing the developer to avoid increases in land value.

Another model could apply to stronger real estate markets. A land bank would acquire land at the current market value, which could be too expensive for it to make sense to hold vacant land while putting together a TOD project or finding an appropriate developer. The land bank would ensure that the property has a cash flow from rents that allows it to hold the property without incurring high carrying costs. The land bank would cover its costs with the rents from the property while it waits for an appropriate redevelopment project. In this scenario, a land bank could still write down the value of the land when transferring it to a developer to make sure that the project remains feasible, especially if the TOD includes affordable or mixed-income housing.

A land bank could also build on traditional land banking to fund infrastructure to support TOD. Under the right market conditions, the assembly and sale or transfer of developable land in station areas could generate enough of an increase in property value that value capture mechanisms could be used to contribute to the cost of infrastructure. This value capture scenario is likely to be applicable only for TOD projects that do not require public-sector assistance. As described above, the value of land is often written down by government entities to make projects feasible. This happens due to the challenges inherent in TOD projects, where a key variable is to make land available to development or reuse at a cost that is low enough to attract some form of private investment capital, even if the development projects are small. Once one key barrier to development (i.e., land costs) is removed, more development projects have the potential to move forward.

EXAMPLES OF LAND BANKING FOR TOD

There are many examples of traditional land banking focused on vacant and abandoned properties (e.g., in Cleveland, Ohio; Detroit, Michigan; and Genesee County, Michigan) and a few examples of land banking for TOD, including those described below (in Denver, Portland, and Dallas), but no examples exist of land banking for TOD infrastructure. Most agencies engaged in property acquisition for TOD are helping others acquire land rather than directly purchasing, owning and maintaining, or banking the properties themselves. Although there are no examples of land banking for TOD infrastructure, helpful lessons can be gleaned from traditional and TOD-focused land banks.

URBAN LAND CONSERVANCY AND LAND BANKING

In Denver, the Urban Land Conservancy (ULC), a nonprofit organization, acquires land using the Denver TOD Fund, a $15-million fund that aims to develop and preserve 1,200 affordable housing units near transit over 10 years. [159]

ULC, which is the only entity approved to borrow from the fund, identifies opportunity sites and takes out short-term loans from the fund to purchase the sites. ULC eventually sells or leases the property to an affordable housing developer when permanent financing becomes available to pay back the loan.

---

[159] The Denver TOD Fund is discussed in Exhibit 22 in Chapter III, Section J.

Part of ULC's mission is to act as a land bank by purchasing and holding opportunity sites in corridors that are slated for future transit development. ULC does not typically purchase vacant or abandoned property; it attempts to acquire sites with an existing revenue stream from rents on the property so that carrying costs are minimal and the property might even generate income.

PORTLAND METRO'S TOD PROGRAM

Portland Metro, the MPO in the Portland, Oregon, region, is the only regional government in the United States that directly acquires and holds land for development as TOD. Since its inception, the TOD Program has acquired properties in several suburban locations around Portland, including in the cities of Milwaukee, Hillsboro, Gresham, and Beaverton. Properties remain in Metro ownership until an appropriate transit-oriented project is proposed. Metro acquired all of these properties opportunistically, as desirable property became available or when the TOD Program had access to federal funding resources that it could use for acquisition (see Exhibit 38).

Metro does not have a policy to purchase only suburban properties, but that is where the opportunities have arisen. Because land values and rents in the suburban locations cannot yet support TOD, Metro's land banking program has often required a relatively long-term investment.

Challenges that Metro has faced include:

- Finding local government partners with interest and experience in real estate development. Without a willing local government partner, a development project cannot go forward. Without that local financial investment, interest and assistance from localities can be limited.

Exhibit 38. Land Banking for TOD Project: The Crossings

In 2001 and 2002, Metro's TOD Program purchased three sites totaling 13 acres around the future Civic Drive MAX light-rail station in Gresham, Oregon. In 2007, The Crossings, a five-story, mixed-use project with 81 homes built above 20,000 square feet of ground-floor retail with below-grade parking, was developed on 1.9 acres of the land. The development team was a public-private partnership between Peak Development, the city of Gresham, Metro, and the state of Oregon.

In addition to purchasing the land, Metro negotiated a disposition and development agreement to ensure a transit-supportive site plan, more housing, and a mix of retail uses. Metro also applied a land value write-down to the project and purchased a TOD easement from the developers to offset the additional construction costs associated with building more compact, mixed-use TOD. These tools required developers to meet certain requirements such as minimum densities, pedestrian-friendly amenities, and reduced parking.

The Crossings, Gresham, Oregon.
Source: © Metro, Portland, Oregon

- Identifying funding to acquire properties, particularly for smaller projects because they can take as long and be as complicated to implement as larger projects but provide fewer benefits.

- Restrictions on the funding source. In at least one instance, Metro used FTA funding to purchase a site, which has added limits to what can be done with the site.

SUSTAINABLE LAND USE DEVELOPMENT FUNDING PROGRAM

In 2006, the North Central Texas Council of Government (NCTCOG) Sustainable Land Use Development Funding Program launched a pilot program to help local governments assemble parcels for redevelopment. The NCTCOG land bank program was not limited to TOD projects. The program loaned $1 million each to four projects, including three TOD projects, one of which has since been canceled.

Under the NCTCOG's land banking pilot program, funding was provided as no-interest loans only to cities, not to private developers. NCTCOG considered working with transit agencies as well, but in Texas transit agencies do not have authority to develop real estate. NCTCOG has faced similar challenges to those faced by Metro when working with local government partners. The NCTCOG land banking program will not condemn properties or pay for relocation costs or the costs associated with a land purchase such as appraisals. Some cities had real estate programs and experienced staff, but other cities were inexperienced in real estate development and had to hire consultants for services such as appraisals. Such costs could be burdensome for the cities.

Negotiating land prices could be tricky because of transparency requirements and public meetings laws. The cities did not want project information to be public because it could affect land prices and so conducted real estate transactions in closed executive sessions, which is allowed under Texas law.

NCTCOG capped the amount to be paid for a property to 110 percent of the appraised value. This limit provided guidance to cities in negotiating deals, since they could tell property owners that they were not permitted to pay more than that amount. The cap of 110 percent also meant that some deals were lost because the owner would not accept the city's maximum offer. NCTCOG found that requiring cities to get approval for individual purchases presented a challenge because of the time it takes to complete the approval process. A parcel can come onto the market and off again very quickly, and a property owner can drop out of a deal if an agency does not move quickly enough. Instead, NCTCOG and the cities identified and approved a zone for land banking, and the city was authorized to purchase any qualifying parcel in the zone.

The land banking program was initially conceived as a revolving loan fund, but none of the original loans have yet been repaid. How the money might be reallocated once the fund is replenished has not been determined. NCTCOG is presently focusing its land banking efforts on new schools, not TOD projects. Because NCTCOG is not charging interest on the funds loaned for land purchases, it receives no income from the program, but it is requiring that cities turn over any profits realized from increases in land values when the land is sold.

LESSONS LEARNED

These examples of land banking for TOD illustrate some of the challenges of this model. For a new model of land banking for TOD infrastructure to work, an agency would need to find or create opportunities to acquire land in strategic locations and obtain the resources to hold and maintain the properties until a suitable TOD project was financially feasible. To do this, a TOD infrastructure land bank would likely need to:

- **Maintain a funding source that is relatively large and flexible.** This is particularly true in a strong real estate market or a location with potential for growth, where the land bank would need sufficient resources and funds to acquire and maintain properties over what is often an indeterminate period of time. Because land banks might need to hold property for a long time until an appropriate development opportunity occurs, significant carrying costs can be involved, including maintaining the property.

- **Think strategically in acquisition and transfer or sale of properties.** Traditional land banking that acquires distressed or vacant properties without regard to their location does not necessarily allow the strategic assembly of parcels into a larger development site. Under a new model of land banking for TOD, the land banking authority would need to use its limited resources carefully to acquire contiguous parcels.

- **Be nimble and flexible.** The real estate market can move very quickly, and government agencies sometimes move too slowly to close deals. The opportunistic nature of property acquisition and the longer financing and development schedules associated with TOD make it especially difficult to secure land and financing within a seller's timeframe. NCTCOG's solution, as described above, was to identify and approve zones for land banking.

- **Tolerate a higher level of risk than is typical of most government agencies.** Regardless of how a property is acquired, acquiring and holding properties can be associated with significant risk and responsibility, depending on local real estate market conditions. For example, vacant or distressed properties could be in a dangerous state of disrepair and require immediate investment to reduce hazards.

- **Find other sources of income so that the land bank does not depend on interest payments or turning a profit.** Funding for traditional land banks typically comes from the land bank's operations (selling other lands) and local government contributions. Because of the challenges inherent in TOD projects, however, land must be available for development at a cost low enough to attract private investment. Reducing land costs eliminates a major barrier to development and could result in more development projects moving forward. However, it means that a TOD land bank might not receive revenue from land sales that could be used for infrastructure projects or other purchases. To address that issue, land banking for TOD could focus on acquiring properties with existing income from rents.

- **Maintain privacy for real estate transactions.** Some land banking for TOD relies on buying land at a lower cost so that future development is more feasible. In this case, a TOD land bank would need to maintain privacy for real estate negotiations to occur without driving up the land values. Some cities have established nonprofit organizations to act on behalf of the city for this purpose.

APPLICABILITY TO OTHER PLACES

This version of land banking would be a complicated model to deploy because it requires the TOD land bank to have sufficient resources and funds to acquire properties and maintain them for an indeterminate time. By carefully sorting out opportunities and focusing efforts in a single location rather than spreading them out over multiple station areas, land banking could be the key to facilitating a catalytic TOD project.

ENTITIES INVOLVED IN IMPLEMENTING THE MODEL

A local government could establish a separate authority with the power to negotiate prices, hold and maintain land, and assist in development. A separate nongovernmental authority might be more suited to land banking for TOD than existing government agencies because it would be able to assemble the real estate expertise necessary for the strategic acquisition and transfer of properties and maintain the privacy necessary for real estate transactions.

Cities, counties, transit authorities, and other public agencies that are involved in the provision of infrastructure must work closely with the land banking authority. There can be tension and unnecessary cost escalation if transit authorities and other government agencies are competing for the same properties.

REFERENCES

Alexander, Frank. *Land Banks and Land Banking*. Center for Community Progress. June 2011. http://www.communityprogress.net/land-bank-book-resources-105.php.

Center for Transit-Oriented Development and Strategic Economics. *CDFIs and Transit-Oriented Development*. Prepared for the Federal Reserve Bank of San Francisco. October 2010. http://www.frbsf.org/publications/community/wpapers/2010/cdfi_transit_oriented_design.html.

Center for Transit-Oriented Development. *Transit-Oriented Development Strategic Plan/Metro TOD Program*. April 2011. http://ctod.org/pdfs/2011PortlandTODweb.pdf.

Metro. "The Crossings." http://www.oregonmetro.gov/index.cfm/go/by.web/id=26409. Accessed February 20, 2012.

Metro. "Transit-Oriented Development." http://www.oregonmetro.gov/tod. Accessed February 20, 2012.

North Central Texas Council of Government. "Sustainable Development Funding Program: Landbanking Projects." http://www.nctcog.org/trans/sustdev/landuse/funding/lbank.asp. Accessed February 20, 2012.

Personal communication with Megan Gibb, Metro TOD Program Manager, by Sarah Graham, Strategic Economics, December 9, 2011.

Personal communication with Karla Weaver, NCTCOG Program Manager, by Sarah Graham, Strategic Economics, December 12, 2011.

Urban Land Conservancy. "Home page." http://www.urbanlandc.org. Accessed February 20, 2012.

D. DISTRICT ENERGY SYSTEMS

Cities across the world are seeking to use energy more efficiently in the built environment to reduce their dependence on fossil fuels. District energy systems have been an effective tool to achieve this goal. There are over 700 district energy systems in the United States in city centers and large campus-based institutions, like hospitals and universities, that benefit from economies of scale.[160] By implementing energy-efficiency improvements as part of a comprehensive, districtwide infrastructure plan in a transit-oriented location, communities could improve the energy efficiency not just of individual buildings, but of the built environment as a whole—reducing energy use of individual buildings, encouraging renewable energy, and facilitating compact development.

District energy involves the production and delivery of steam, hot water, or chilled water from a centralized plant or mini-plants to multiple buildings via an underground pipeline system.[161] It is a reliable, efficient, cost-effective way to provide climate control without onsite boilers, chillers, or air conditioners.[162] To deliver district energy services, a utility service provider assumes responsibility for capital investments, generates (or captures) and delivers energy, and charges building owners for use of the system.[163] District energy systems typically rely on combined heat and power, also known as cogeneration, which is the simultaneous production of electricity and heat from a single fuel source. In a conventional power plant only approximately one-third of the energy consumed is converted to electricity, and the remainder is lost as heat. Combined heat and power captures some of this lost energy by using the heat to provide heating to the power plant or to buildings that are connected to the power plant through the pipe network.[164]

District energy systems are not a financing or funding mechanism for TOD but rather a development approach that could help cities and building owners use cleaner energy sources. By building a district energy system at the same time as other districtwide infrastructure improvements, communities could meet multiple environmental goals, leverage financing sources, and reduce overall construction costs. By pooling projects for financing purposes, a city could issue fewer, larger bonds, reducing transaction costs, and combine multiple types of grants (e.g., for energy efficiency, streetscape, sewer, and other improvements). Implementing multiple improvements at once could also lower construction costs and disruptions—for example, by reducing the number of times that a city tears up a street.

---

[160] Environmental and Energy Study Institute. "What is District Energy?" 2011.

[161] Energy Systems. "District Heating & Cooling: Frequently Asked Questions." http://www.esc-omaha.com/heating-and-cooling/faqs.aspx. Accessed August 20, 2012.

[162] International District Energy Association. "What is District Energy?" http://districtenergy.org/what-is-district-energy. Accessed July 24, 2012.

[163] National Trust for Historic Preservation. *The Role of District Energy in Greening Existing Neighborhoods.* 2010.

[164] Environmental and Energy Study Institute. "Renewable Energy." http://www.eesi.org/renewable_energy?page=8. Accessed August 20, 2012.

THE ROLE OF DISTRICT ENERGY SYSTEMS IN FUNDING AND FINANCING TOD INFRASTRUCTURE

District energy systems could be helpful in revitalizing communities and encouraging more sustainable and compact growth. Some of the benefits that district energy systems could offer include:

- **Reduce construction costs of new development.** Because district energy systems are centralized systems, buildings do not need onsite boilers, chillers, or air conditioners, which lowers construction costs because less square footage and equipment is needed, particularly for commercial and office buildings. In areas where the district energy system is already in place, redevelopment could be more cost-effective, which could encourage development around transit stations.

- **Encourage compact growth.** For district energy to be viable, buildings must be close enough to each other to take advantage of economies of scale. Additionally, since district energy systems are more efficient when they have more diverse users, they encourage a mix of uses. For example, residences consume more energy during the morning and at night, while commercial land uses consume more energy during the day.

- **Reduce greenhouse gas emissions and other pollution.** By using renewable energy, communities could reduce pollution from power generation, including the greenhouse gases that contribute to climate change.

- **Improve energy supply reliability.** Most district energy systems operate at a reliability of well over 99 percent with virtually no interruptions in service.[165] District energy systems have their own energy source rather than relying on the electricity grid. Most district energy users in the United States are campus-based institutions like hospitals of the importance of reliability to their operations.

- **Promote economic development.** District energy lowers operating costs of energy-intensive industries, which makes the community more appealing to businesses. Over the long run, this could strengthen the local real estate market and spur demand for TOD in a downtown or transit district with a district energy system.

- **Reduce buildings' maintenance and operation costs.** District energy systems deliver less-expensive energy through improved efficiency and economies of scale. Savings come from reduced building operations and maintenance costs since no chillers or heaters are needed. However, the savings take time to accumulate. Generally, it takes eight to 10 years for a building owner to recover the initial investment in a district energy system.

---

[165] International District Energy Association. "What is District Energy?" http://districtenergy.org/what-is-district-energy. Accessed February 20, 2012.

WEST UNION, IOWA, GREEN PILOT PROJECT

West Union, a small town in northeast Iowa with 2,500 residents, has an ambitious plan to revitalize its downtown, and improved energy efficiency is a key part of that plan. The town is developing a district energy system based on geothermal energy in a neighborhood of historic buildings.[166]

West Union's Main Street district comprises 60 buildings and a total floor area of 330,000 square feet. Most of the buildings' existing heating and cooling systems require updating for connection to the district energy system. The cost and required updates will vary depending in part on the age of the systems.[167] The external system components will be funded by multiple public-sector sources, and property owners will be responsible for funding the internal building improvements necessary to use the district energy system. However, property owners do not have to use the network. Each owner is free to join only if and when it makes financial sense to do so. To encourage property owners to join the system, the city is contemplating offering a one-time incentive to building owners.

West Union is using a combination of federal, state, and local funds (Exhibit 39). This project also benefits from coordination with a complete streets project, which is rebuilding a street to make it safer and more appealing to pedestrians and bicyclists.

Funder	Amount
Department of Energy—Energy Efficiency Conservation Block Grant	$837,500
EPA Climate Change Showcase Communities Grant	$500,000
Fayette County	$10,000
I-JOBS*	$1,175,000
Iowa Department of Agriculture & Land Stewardship—I-JOBS*	$500,000
Iowa Department of Cultural Affairs—Iowa Great Places	$160,000
Iowa Department of Natural Resources—I-JOBS*	$100,000
Iowa Department of Transportation—Revitalize Iowa's Sound Economy	$2,327,034
Iowa Economic Development—CDBG	$1,000,000
Iowa Economic Development—Sustainable Communities Demonstration	$229,000
Iowa Watershed Improvement Review Board	$500,000
Main Street West Union	$10,995
U.S. Department of Agriculture—Rural Community Development Initiative	$37,000
Main Street Iowa Challenge Grant	$100,000
Main Street Iowa—I-JOBS*	$440,000
West Union, City of	$2,368,499
Total	**$10,295,028**

Exhibit 39. Funding sources.

*I-JOBS is a state infrastructure investment program paid for by issuing construction bonds.
Source: Iowa Economic Development. "West Union Green Pilot Project
Partners." http://www.iowaeconomicdevelopment.com/community/westunion/partners.aspx. Accessed July 24, 2012.

---

[166] National Trust for Historic Preservation. *District Energy in West Union, IA: Integrating a New District Energy System into a Historic Main Street Community*. 2010.

[167] Ibid.

Securing grants and outside funding meant West Union did not have to tax property owners through a special assessment district.[168] Other cities or communities interested in implementing district energy systems could establish a special assessment district to finance it.

LESSONS LEARNED

- **District energy is becoming relevant to neighborhood revitalization efforts by helping communities achieve broader environmental goals while encouraging economic development.** The main incentives for building owners to become part of a district energy system are reliability and long-term net operating cost savings. However, district energy systems bring additional benefits that support TOD, including making infill redevelopment less expensive, encouraging compact and mixed-use development, and making communities cleaner and healthier.

- **The upfront costs of implementing district energy tend to be onerous.** Using state or federal grants if available makes district energy more attractive and encourages property owners to connect to the network. Another way to finance district energy is through a special assessment district. These districts generally require voter approval, so the government agency interested in implementing the district would have to educate voters about the potential benefits and how they compare to the costs.

- **A critical mass is necessary to take advantage of economies of scale.** Locations suitable for district energy are typically dense and compact neighborhoods with a stable or growing demand for heating and cooling services, like many TOD districts. District energy could probably not be implemented in weak real estate markets where occupancy rates are low and demand for energy—and therefore the revenue stream to pay the system operator—is unreliable.

- **Various ownership structures have been used for district energy projects.** To establish the appropriate structure for a project, local governments can consider local opportunities and constraints, including opportunities for partnerships among private- and public-sector and nonprofit entities. The optimal structure will enable the project to take advantage of low-cost financing, access available grants and incentives, pursue favorable tax treatment, and help facilitate the most efficient and effective transfer of risk to project partners.

- **Establishing partnerships among public entities and between the public and private sectors is critical to implementing district energy.** District energy can be part of a comprehensive approach to environmental and economic sustainability in which local governments, state agencies, and building owners cooperate to reach common goals.

APPLICABILITY TO OTHER PLACES

Many transit-oriented districts could be good candidates for district energy because they have compact, mixed-use development. A municipality would take several steps to implement district energy systems, including:

---

[168] Ibid.

- **Encourage the use of energy-efficient technologies through energy city or regional policies** such as the establishment of goals for reducing energy demand or generating clean energy.[169]

- **Take advantage of economies of scale by implementing a comprehensive approach to development.** District energy is a promising tool for communities seeking a comprehensive approach to environmental sustainability. By implementing several complementary projects at once, such as district energy and TOD streetscape infrastructure improvements, cities can reduce overall construction costs. Implementing district energy requires financing sources, which typically involves access to debt markets through issuing bonds (public or private) or loans. Since the cost of streetscape improvements is marginal compared to the total cost of a district energy system, the local government could group district energy and street projects together and get financing through the bond market.

- **Identify areas that are likely to successfully implement district energy.** As a rough rule of thumb, a neighborhood will be a good candidate if it has some of the following characteristics: several large buildings or building complexes (e.g., hospitals, hotels, or colleges), a mix of uses as in a town or village center, relatively high residential densities (e.g., multifamily units or apartments), relatively little space between buildings, a street grid to make the layout of the system more efficient, a source of relatively cheap energy (e.g., waste heat from a boiler or sewage treatment facility), and few electric resistance heating systems, which convert nearly all of the energy in electricity to heat, but cannot be easily retrofitted.

- **Build institutional capacity.** District energy systems are challenging to develop and will require significant staff time and commitment.[170]

- **Secure the customer base.** District energy systems are viable only with enough building owners in the area who are interested in long-term energy reliability.[171]

ENTITIES INVOLVED IN IMPLEMENTING THE MODEL

Leadership can come from the public or private sectors. Local governments can encourage building owners to switch to more efficient technologies by implementing policies that support these technologies such as regional targets for energy demand reduction and funding assistance programs.

Potential users of the system are also key since they will be required to enter into a long-term agreement with the district to purchase its energy. Therefore, understanding their needs and constraints is essential for successful implementation of a district energy system.

REFERENCES

Energy Systems. "District Heating & Cooling: Frequently Asked Questions." http://www.esc-omaha.com/heating-and-cooling/faqs.aspx. Accessed August 20, 2012.

---

[169] National Trust for Historic Preservation. *The Role of District Energy in Greening Existing Neighborhoods.* 2010.

[170] Ibid.

[171] Ibid.

Environmental and Energy Study Institute. "Renewable Energy." http://www.eesi.org/renewable_energy?page=8. Accessed August 20, 2012.

Environmental and Energy Study Institute. "What is District Energy?" 2011. http://www.eesi.org/district_energy_092311.

International District Energy Association. "What is District Energy?" http://districtenergy.org/what-is-district-energy. Accessed February 20, 2012.

Iowa Economic Development. "West Union Green Pilot Project." http://www.iowaeconomicdevelopment.com/community/westunion/energy.aspx. Accessed February 20, 2012.

Iowa Economic Development. "West Union Green Pilot Project Partners." http://www.iowaeconomicdevelopment.com/community/westunion/partners.aspx. Accessed July 24, 2012.

National Trust for Historic Preservation. *District Energy in West Union, IA: Integrating a New District Energy System into a Historic Main Street Community*. 2010. http://www.preservationnation.org/information-center/sustainable-communities/sustainability/green-lab/additional-resources/West-Union_FINAL.pdf

National Trust for Historic Preservation. *The Role of District Energy in Greening Existing Neighborhoods*. 2010. http://www.preservationnation.org/information-center/sustainable-communities/sustainability/green-lab/additional-resources/District-Energy-Long-Paper.pdf.

APPENDIX A. EPA SMART GROWTH IMPLEMENTATION ASSISTANCE PROGRAM

Communities around the country are looking to get the most from new development and to maximize their investments. Frustrated by development that gives residents no choice but to drive long distances between jobs and housing, many communities are bringing workplaces, homes, and services closer together. Communities are examining and changing zoning codes that make it impossible to build neighborhoods with a variety of housing types. They are questioning the fiscal wisdom of neglecting existing infrastructure while expanding new sewers, roads, and services into the fringe. Many places that have been successful in ensuring that development improves their community, economy, and environment have used smart growth principles to do so (see box). Smart growth describes development patterns that create attractive, distinctive, and walkable communities that give people of varying age, wealth, and physical ability a range of safe, convenient choices in where they live and how they get around. Growing smart also means that we use our existing resources efficiently and preserve the lands, buildings, and environmental features that shape our neighborhoods, towns, and cities.

However, communities often need additional tools, resources, or information to achieve these goals. In response to this need, the Environmental Protection Agency (EPA) launched the Smart Growth Implementation Assistance (SGIA) program to provide technical assistance—through contractor services—to selected communities.

The goals of this assistance are to improve the overall climate for infill, brownfields redevelopment, and the revitalization of non-brownfield sites—as well as to promote development that meets economic, community, public health, and environmental goals. EPA and its contractor assemble teams whose members have expertise that meets community needs. While engaging community participants on their aspirations for development, the team can bring their experiences from working in other parts of the country to provide best practices for the community to consider.

Since 2009, EPA has engaged staff from the DOT and HUD in SGIA projects. This collaboration is part of the HUD-DOT-EPA Partnership for Sustainable Communities, under which the three agencies work together to help improve access to affordable housing, more transportation options, and lower transportation costs while protecting the environment in communities nationwide. Using a set of guiding livability principles and a partnership agreement, this partnership

---

Smart Growth Principles

Based on the experience of communities around the nation, the Smart Growth Network developed a set of ten basic principles:

1. Mix land uses.

2. Take advantage of compact building design.

3. Create a range of housing opportunities and choices.

4. Create walkable neighborhoods.

5. Foster distinctive, attractive communities with a strong sense of place.

6. Preserve open space, farmland, natural beauty, and critical environmental areas.

7. Strengthen and direct development towards existing communities.

8. Provide a variety of transportation choices.

9. Make development decisions predictable, fair, and cost effective.

10. Encourage community and stakeholder collaboration in development decisions.

Source: Smart Growth Network. "Why Smart Growth?" http://www.smartgrowth.org/why.php.

coordinates federal housing, transportation, and other infrastructure investments to protect the environment, promote equitable development, and help to address the challenges of climate change.

For more information on the SGIA program, including reports from communities that have received assistance, see *www.epa.gov/smartgrowth/sgia.htm*.

For more information on the Partnership for Sustainable Communities, see *www.sustainablecommunities.gov*.

APPENDIX B. TOOLS FOR FUNDING AND FINANCING TRANSIT ORIENTED DEVELOPMENT INFRASTRUCTURE

This appendix gives more information on 30 of the existing and emerging tools for funding and financing[1] the infrastructure needed to support transit-oriented development (TOD) that were introduced in Chapter II. Like the report as a whole, this appendix focuses on funding and financing tools for the capital costs associated with TOD-related infrastructure (including sewer, water, storm drain, and other utilities; roads; bicycle and pedestrian improvements; parks; streetscape improvements; and structured parking) rather than on funding for operations and maintenance of that infrastructure. However, in some cases the tool can apply to operations and maintenance as well as capital uses.

The tools apply across a broad range of places, including existing and planned station areas and transit corridors, and across a variety of market contexts, ranging from strong markets with significant development activity to weaker markets where there may be little or no demand for available land. Not all tools work well in all contexts, and a key task in creating a TOD infrastructure financing strategy is to evaluate which tools will work best in a given context. While this appendix describes tools individually, most TOD infrastructure financing strategies combine multiple tools. Note also that many of the tools involve highly complex financial transactions requiring legal and financial expertise.

KEY FACTORS FOR EVALUATING TOOLS

The tool profiles are organized around the factors that a local government might consider in determining whether the tool is appropriate for its situation. The tool profiles include:

- **Applicability to different types of infrastructure:** The most typical uses for the tool, as well as other allowable uses, with a focus on TOD infrastructure as defined in this report.

- **Approval requirements and legal and political considerations:** What it takes to get the tool approved for use, including whether the tool requires voter approval or is accessed through a competitive process.

- **Application for strong and weak real estate markets:** The extent to which implementation of the tool relies on local real estate market conditions.

- **Capacity and scale:** What size or scale of project the tool can be used for and/or factors that determine the amount of funding that the tool can generate. Some of the tools can be used only for projects that meet certain cost thresholds or are typically used for projects that fall within a range of costs; these thresholds and ranges are noted where they apply. In general, however, few rules of thumb apply for determining how large or small a project must be for a tool to be applicable. Instead, communities must consider whether a project is of sufficient size to justify the transaction costs involved in accessing a given funding source. Depending on the tool, those costs could include writing a grant application or structuring a complex financial transaction. The applicability of many

---

[1] As discussed in Chapter II, "funding" refers to a revenue stream or source of revenue; "financing" refers to the mechanisms used to manipulate available revenue streams so that agencies can provide infrastructure before revenue equal to the full cost of that infrastructure becomes available.

tools to any particular project also depends on the extent to which a state, regional, or local government prioritizes resources (e.g., federal block grants or bonding capacity) for TOD.

- **Ease of use:** The ease of implementing and administering the tool.

- **Timing and lifecycle:** The terms of the financing and any specific repayment structures, including credits and reimbursements, necessary under the tool.

In addition to these key factors for evaluation, the tool profiles also describe:

- **Other limitations of the tool:** Any other restrictions on how the tool can be used.

- **Use of the tool in practice:** How widely the tool has been used to fund or finance TOD-related infrastructure, including an example where possible. Some of the tools have the potential to be used in a TOD context but have not yet been used for TOD-related infrastructure; in these cases, examples were chosen that illustrate how the tool is typically used. [2]

- **Getting started:** How a local government (or, in some cases, another entity) might begin to implement the tool in its jurisdiction.

Where relevant, the profiles also discuss the tools' application to land assembly, new versus existing station areas, and operations and maintenance; risks involved in implementing the tool; and sources of capital to pay back the financing, if required.

TOOLS

The tools are organized into seven categories:

A. **Direct fees:** Charges paid by the users of the infrastructure.

 1. User fees and transportation utility fees

 2. Congestion pricing

B. **Debt tools:** Mechanisms for borrowing money to finance infrastructure.

 1. Industrial loan companies and industrial banks

 2. General obligation bonds

 3. Revenue bonds

 4. Private activity bonds

 5. Certificates of participation and lease revenue bonds

 6. Revolving loan funds

 7. State infrastructure banks

---

[2] The examples are intended to be geographically diverse. However, in many cases the only or best available example was in California.

8. Grant anticipation revenue vehicle bonds

9. Railroad Rehabilitation and Improvement Financing

C. **Credit assistance:** Mechanisms that improve the creditworthiness of the borrower issuing a bond or requesting a loan and thus provide access to better borrowing terms.

    1. Credit assistance tools

    2. Transportation Infrastructure Finance and Innovation Act

D. **Equity sources:** Tools that allow private entities to invest (i.e., take an ownership stake) in infrastructure in expectation of a return.

    1. Public-private partnerships

    2. Infrastructure investment funds

E. **Value capture mechanisms:** Tools that capture the increased value or savings resulting from the public provision of new infrastructure.

    1. Developer fees and exactions

    2. Special districts

    3. Tax increment financing

    4. Joint development

F. **Grants:** Funds that do not need to be paid back.[3]

    1. Congestion Mitigation and Air Quality Improvement Program

    2. Transportation Alternatives Program (Formerly Transportation Enhancements Program)

    3. Urbanized Area Formula Funding Program

    4. Community Development Block Grant Program

    5. Economic Development Administration grants

    6. Foundation grants

    7. Program-related investments

G. **Emerging tools:** New concepts for making TOD-related infrastructure possible. Most of the tools in this category do not fit neatly into any of the other categories.

    1. Structured funds

    2. Land banks

    3. Redfields to greenfields

    4. National infrastructure bank

---

[3] Many types of grants are available at the state and regional levels that are not included in this report because they vary by location.

A. DIRECT FEES

A-1. USER FEES AND TRANSPORTATION UTILITY FEES

User fees include the fees charged for the use of public infrastructure or goods (e.g., a toll road or bridge, water or wastewater systems, or public transit). Fees are typically set to cover (or partially cover) a system's operating and capital expenses each year, which can include debt service for improvements to the system.

Transportation utility fees are assessments on property that are designed to be closely related to transportation demand and can therefore spread the costs of financing local roads or other transportation services among users in a fashion that approximates a user fee. Because it is not a tax, a transportation utility fee typically does not require voter approval.[4] The fee can be a flat fee for each property, or it can apply a formula based on units of housing, number of parking spaces, or square footage. It can also be based on the estimated trip generation rate for a property type. Transportation utility fees are most commonly used for roads, but they can also be used to provide a dedicated funding source for transit systems.

KEY FACTORS FOR EVALUATION

Applicability to different types of infrastructure: User fees are commonly associated with a variety of infrastructure types, including transit, parking, water and wastewater systems, and toll roads and bridges.

Approval requirements and legal and political considerations: Fees for public services or facilities are typically limited to the actual costs of providing the service or facility. Implementing a new fee or raising an existing fee typically requires local legislative action but not voter approval. However, there are political considerations in implementing new fees or raising existing ones, as public opinion is relevant in making these decisions.

Application for strong and weak real estate markets: This tool does not rely on new development and is therefore applicable in strong and weak real estate markets. However, the amount of revenue that can be raised through fees generally depends on local conditions. For example, parking fees can be set higher in places with strong parking demand and a limited supply (e.g., downtowns).

Capacity and scale: The revenue from user fees can help offset operations and maintenance costs or help finance new infrastructure. The scale of infrastructure that can be financed depends on the size of the fee and the size of the base (i.e., the number of users who pay the fee). To cover the cost of building infrastructure in addition to operations and maintenance, user fees might need to be raised.

Ease of use: Depending on the service or facility provided, implementing user fees on publicly owned infrastructure could require legislative approval, technical and financial feasibility studies, and

---

[4] Lari, A. et al. *Value Capture for Transportation Finance: Technical Research Report.* Center for Transportation Studies, University of Minnesota. 2009.

environmental clearance. For example, Denver's Regional Transit District's ability to set parking fees is restricted by state legislation.

USE OF THE TOOL IN PRACTICE

User fees are widely used to pay for operations; maintenance; and construction of transit, parking, and other rate-based infrastructure or utilities (including sewer, water, and toll roads and bridges).

Example: Corvallis Sustainability Initiatives Fee (Transportation Utility Fee)

Location: Corvallis, Oregon (Exhibit B-1)

Description: In February 2011, the city of Corvallis implemented a sustainability initiatives fee that pays for free bus service and maintenance of sidewalks and public trees. The city charges residents and businesses the fee via the water bill to reduce administrative costs. The transit portion of the fee varies by the number of trips a property is expected to generate; a single-family residential property is charged $2.75 a month. In addition, all properties are charged $1.30 a month for sidewalk and public tree maintenance. The sidewalk maintenance and urban forestry fees are based on the assumption that all residents and businesses benefit equally from sidewalks and from a healthy public tree system.[5]

Exhibit B-1. Downtown Corvallis.
Source: Wendell Ward via Flickr.com

GETTING STARTED

The process for establishing a user fee for publicly owned infrastructure depends on ownership of the asset but is typically led by the local government. The process includes assessing:

- The project's technical feasibility to determine whether the project could be completed and to assess users' willingness to pay.

- The project's financial feasibility to determine if the project will cover its construction and operation costs.

- Legislative requirements to determine whether existing state and local legislation allows implementation of new fees.

---

[5] City of Corvallis. "Sustainability Initiatives Funding." http://www.ci.corvallis.or.us/index.php?option=com_content&task=view&id=3973&Itemid=4490. Accessed December 9, 2011.

REFERENCES

City of Corvallis. "Sustainability Initiatives
Funding." http://www.ci.corvallis.or.us/index.php?option=com_content&task=view&id=3973&Itemid=4490.
Accessed December 9, 2011.

Lari, Adeel et al. *Value Capture for Transportation Finance: Technical Research Report.* Center for Transportation
Studies, University of Minnesota.
2009. http://www.cts.umn.edu/Publications/ResearchReports/reportdetail.html?id=1802.

A-2.   CONGESTION PRICING

Congestion pricing is the use of pricing mechanisms to manage demand for services during peak periods. The economic rationale is that, at a price of zero, demand exceeds supply, causing a shortage, and that the shortage could be corrected by charging a price rather than by increasing the supply. Usually this means increasing prices in certain times or places where congestion occurs or introducing a new user fee when peak demand exceeds available supply. Congestion pricing has been widely used by telephone and electric utilities and public transit agencies.[6] More recently, it has been implemented to mitigate congestion on roadways and bridges.

Examples of congestion pricing include the I-15 High Occupancy Toll (HOT) lanes in San Diego, California; SR-167 in Seattle, Washington; I-25 in Denver, Colorado; and I-394 in Minneapolis, Minnesota. In San Diego, the I-15 HOT lanes tolls adjust dynamically based on real-time traffic demand. Fees adjust in 25-cent increments as often as every six minutes to help maintain free-flowing traffic in the HOT lanes.[7] In 1998, the Midpoint and Cape Coral bridges in Lee County, Florida, implemented variable pricing. Since 2003 London uses congestion pricing in its central business district, charging US$16 for some categories of motor vehicles to travel within a certain zone between 7 a.m. and 6 p.m. Monday through Friday. The goal of the fee is to reduce congestion and raise funds for transportation improvements.[8]

KEY FACTORS FOR EVALUATION

Applicability to different types of infrastructure: The revenue from congestion pricing initiatives has been generally allocated to both the cost of the tolling system and improving the highway and/or transit system. The uses vary depending on the public authorities' needs and priorities. In San Diego, half of the annual revenue of the I-15 HOT lanes is used to support transit service in the I-15 corridor. In Minnesota, state law specifies that the revenue collected should first be used to cover the operation of the tolling system, while remaining revenue can be used to support capital improvements for roads or transit. However, no excess revenue has been generated so far to support capital improvements.[9]

Approval requirements and legal and political considerations: The implementation of congestion pricing typically requires state legislation and, in some cases, voter approval. Public opinion is generally a factor to its implementation.

Application for strong and weak real estate markets: Congestion pricing is not directly affected by weak or strong real estate markets. However, real estate markets are generally correlated with economic activity, which in turn is correlated with travel demand. During economic downturns, people tend to make fewer trips than during boom years, and therefore toll revenue can drop.

---

[6] The World Bank. *Sustainable Transport: Priorities for Policy Reform.* Washington, D.C. 1996. pp. 48-49.

[7] FHWA. "Road Pricing Defined." http://www.fhwa.dot.gov/ipd/revenue/road_pricing/defined/demand_mgmt_tool.htm. Accessed August 20, 2012.

[8] Eltis. "An integrated approach to implementing Congestion Charging in London, England." http://www.eltis.org/index.php?id=13&study_id=3062. Accessed August 20, 2012.

[9] Minnesota Department of Transportation. "MnPass Express Lanes." http://www.mnpass.org/394/index.html. Accessed July 23, 2012.

Capacity and scale: Revenue collected from tolls varies with demand, the toll rate, and availability of competing free routes or alternative modes of transportation. Revenue could vary from a few million dollars annually to hundreds of millions of dollars depending on the project. The San Francisco–Oakland Bay Bridge generates approximately $160 million a year.

Ease of use: Using congestion pricing to raise revenue requires legislative approval, technical and financial feasibility studies, and environmental clearance. Though congestion pricing can be an effective tool to relieve congestion and support development of transit-related projects, the process can be time consuming and demands political commitment.

USE OF THE TOOL IN PRACTICE

Several places have implemented congestion pricing on roads by charging different prices during peak hours.

Example: SR-91 Express Lanes

Location: Orange County, California

Description: SR-91 in Orange County has used congestion pricing since the mid-1990s. As required by state law, the Orange County Transportation Authority, in consultation with the California Department of Transportation and the Riverside County Transportation Commission, annually issues the SR-91 Implementation Plan. The plan establishes a multiphase program of projects eligible for funding from excess SR-91 Express Lanes toll revenue, which can include transit and highway improvements. In 2010, SR-91 generated nearly $42 million in revenue. The first set of projects is anticipated to be completed by 2016 and includes six improvements at a total cost of approximately $1.57 billion. [10]

Example: San Francisco's Congestion Pricing

Location: San Francisco, California

Description: In 2004, the San Francisco County Transportation Authority began exploring the possibility of introducing congestion pricing in the downtown area, motivated by the initial success of the London congestion charge. Since then, the transportation authority has studied several proposals that could generate $60 to $80 million of annual revenue for public transit improvement projects and pedestrian and bike infrastructure enhancements. If San Francisco decides to implement congestion pricing, it would likely begin after 2015. [11]

GETTING STARTED

The process for a state or local authorities to get started includes:

---

[10] Orange County Transportation Authority. "91 Express Lanes." http://www.octa.net/91overview.aspx. Accessed September 30, 2011.

[11] San Francisco County Transportation Authority. *San Francisco Mobility, Access, and Pricing Study*. December 2010.

B-8

- Analyzing the project's technical feasibility to determine whether travel demand and users' willingness to pay would support the system and whether suitable technology is available.

- Analyzing its financial feasibility to determine if the project will be able to cover its construction and operation costs, if any excess funds might be generated, and, if so, how the excess could be allocated to other purposes.

- Assessing legislative requirements to determine whether existing state and local legislation allows congestion pricing.

- Getting the necessary federal and perhaps state environmental clearance.

REFERENCES

Eltis. "An integrated approach to implementing Congestion Charging in London, England." http://www.eltis.org/index.php?id=13&study_id=3062. Accessed August 20, 2012.

FHWA. "Road Pricing Defined." http://www.fhwa.dot.gov/ipd/revenue/road_pricing/defined/demand_mgmt_tool.htm. Accessed August 20, 2012.

Minnesota Department of Transportation. "MnPass Express Lanes." http://www.mnpass.org/394/index.html. Accessed July 23, 2012.

Orange County Transportation Authority. "91 Express Lanes." http://www.octa.net/91overview.aspx. Accessed September 30, 2011.

San Francisco County Transportation Authority. *San Francisco Mobility, Access, and Pricing Study*. December 2010. http://www.sfcta.org/images/stories/Planning/CongestionPricingFeasibilityStudy/PDFs/MAPS_study_final_lo_res.pdf.

The World Bank. *Sustainable Transport: Priorities for Policy Reform*. Washington, D.C. 1996.

B. DEBT TOOLS

B-1. INDUSTRIAL LOAN COMPANIES AND INDUSTRIAL BANKS

An industrial loan company (ILC) or industrial bank (IB) is a state-chartered institution with banking powers. ILCs and IBs are regulated by their state chartering authorities and, at the federal level, by the Federal Deposit Insurance Corporation. This type of financial institution can be owned by a non-financial institution, such as a private or publicly held company not typically associated with banking activities (e.g., General Electric, General Motors, Merrill Lynch, Morgan Stanley, American Express, Target, Nordstrom, and Harley-Davidson). As of 2011 ILCs and IBs are permitted in seven states: California, Colorado, Minnesota, Indiana, Hawaii, Nevada, and Utah. Most are in California, Nevada, and Utah.

Although ILCs and IBs are regulated at the state level and regulations vary by state, they are largely subject to the same regulatory and supervisory oversight as commercial banks. In some states, however, ILCs and IBs are not subject to the same usury limits as retail banks and can charge higher interest rates on credit cards. [12] Because ILCs and IBs are not as restricted in some ways as other types of financial institutions, they have been an attractive way for corporations to enter the financial services market, and many major corporations have established IBs.

However, the differences from retail banks have also made ILCs and IBs controversial. In 2011, the Dodd-Frank Wall Street Reform and Consumer Protection Act extended a moratorium on new charters for commercially owned ILCs that had originally been put in place for all ILCs in 2006. The legislation also directed the U.S. Government Accountability Office to study whether ILCs and IBs pose any threat to the stability of the financial system. The GAO's report included various views on this issue. [13]

Most ILCs and IBs serve as small financing companies; however, some have expanded their operations to include some commercial and collateralized real estate lending. Largely operating as a bank would, ILCs and IBs are authorized to make consumer and commercial loans, issue credit cards, and to accept federally insured deposits, but most ILCs and IBs do not operate on a retail basis with branches in a community. ILCs and IBs are subject to the Community Reinvestment Act, a federal law designed to encourage banks to help meet the credit needs of the communities in which they operate, including low- and moderate-income neighborhoods. [14]

Because ILCs and IBs do not typically lend to individuals and small businesses in the community, they must find other ways to meet the CRA requirements. Methods include buying housing bonds issued by government agencies and making CRA-qualified loans for a public purpose. [15] In Utah, for example, UBS Bank USA, an ILC that is the banking affiliate of UBS Wealth Management Americas, is providing the

---

[12] U.S. Government Accountability Office. *Industrial Loan Companies, Recent Asset Growth and Commercial Interest Highlight Differences in Regulatory Authority.* 2005. p. 22.

[13] U.S. Government Accountability Office. *Characteristics and Regulation of Exempt Institutions and the Implications of Removing the Exemptions.* 2012.

[14] Federal Financial Institutions Examination Council. "Community Reinvestment Act: Background and Purpose." http://www.ffiec.gov/CRA/history.htm. Accessed February 21, 2012.

[15] Personal communication with Darryl Rude, Utah Department of Financial Institutions, by Sarah Graham, Strategic Economics, on February 21, 2012.

Utah Housing Corporation, a public corporation created by the state, with a $150-million line of credit to help ensure the housing corporation's ability to provide mortgages and down-payment assistance for first-time homebuyers. The Utah Housing Corporation typically finances mortgages through the sale of tax-exempt housing bonds. Because the sale of bonds might occur infrequently, the UBS-backed line of credit fills a short-term financing gap for the housing corporation.

Since ILCs and IBs operate largely as commercial banks do, financing for TOD infrastructure from an ILC or IB would typically be like getting financing from any other bank, and government agencies would be subject to the same interest rates as any conventional loan. Government agencies might be able to establish agreements with ILCs or IBs for TOD infrastructure loans that have better interest rates and terms if the loan helps ILCs and IBs meet their CRA requirements.[16]

KEY FACTORS FOR EVALUATION

Applicability to different types of infrastructure: Because IBs issue loans that must be repaid, they appear to be most applicable to infrastructure types that generate revenue, such as sewer or water projects that charge a user fee. However, IBs appear to be largely untested in financing infrastructure.

Approval requirements and legal and political considerations: Borrowing from an IB would likely be similar to borrowing from a commercial bank and would not involve significant approval or legal considerations.

Application for strong and weak real estate markets: This financing tool does not rely on new development and therefore has the potential to be applicable in both strong and weak real estate markets.

Capacity and scale: The scale of projects that could be financed with this tool would depend on the size of the IB and its sources of capital for lending.

Ease of use: As described under approval requirements, establishing a new IB is permitted in only seven states, but institutions authorized in those states can offer services nationwide. More widespread use of ILCs and IBs could require enabling legislation in the state.

OTHER LIMITATIONS OF THE TOOL

Industrial loan companies or industrial banks are not permitted in 43 states. Loans are limited to purchasing or refinancing commercial property, mixed-use properties, offices, retail, industrial and warehouse buildings, manufacturing plants, personal loans, consumer loans, and mortgages.

USE OF THE TOOL IN PRACTICE

This tool appears untested for infrastructure.

---

[16] Ibid.

GETTING STARTED

Borrowing from an industrial loan company or industrial bank would likely be similar to borrowing from a commercial bank; a local government would approach the ILC or IB directly to inquire about the availability of loans.

REFERENCES

Federal Financial Institutions Examination Council. "Community Reinvestment Act: Background and Purpose." http://www.ffiec.gov/CRA/history.htm. Accessed February 21, 2012.

Personal communication with Darryl Rude, Utah Department of Financial Institutions, by Sarah Graham, Strategic Economics, on February 21, 2012.

U.S. Government Accountability Office. *Characteristics and Regulation of Exempt Institutions and the Implications of Removing the Exemptions.* 2012. http://www.gao.gov/products/GAO-12-160.

U.S. Government Accountability Office. *Industrial Loan Companies, Recent Asset Growth and Commercial Interest Highlight Differences in Regulatory Authority.* 2005. http://www.gao.gov/products/GAO-05-621.

B-2. GENERAL OBLIGATION BONDS

General obligation bonds are a type of municipal bond[17] used in general public finance or municipal finance. General obligation bonds are generally tax-exempt and are issued for municipal projects that do not generate revenue. They are backed by the "full faith and credit" of the issuer rather than the revenue from a project.[18] Typically, no assets are used as collateral.

General obligation bonds can be issued by government entities including states, counties, cities, redevelopment agencies, special-purpose districts, school districts, or public utility districts. The issuer uses proceeds from the bond sale to pay for capital projects, such as utilities, housing, public transit facilities, parks, water delivery systems, and other projects, or for other purposes that it cannot or is not willing to pay for with other available funds.[19]

KEY FACTORS FOR EVALUATION

Applicability to different types of infrastructure: The proceeds from a general obligation bond sale can pay for TOD capital projects, such as housing, transit facilities, parks, or street improvements. Tax regulations allow certain exceptions and allow using general obligation bonds to fund other items such ongoing operations and maintenance expenses.

Approval requirements and legal and political considerations: State law generally sets the requirements and conditions for issuing general obligation bonds. In some cases, the bond issuance might require voter approval.

Application for strong and weak real estate markets: General obligation bond issuance can be affected by real estate market conditions if part of the revenue stream that serves the debt comes from real estate taxes. In a weak real estate market, investors might require credit assistance (e.g., debt reserves) to prevent default or impose more stringent debt terms (e.g., shorter maturity or higher debt service coverage ratios).

Capacity and scale: There is no limit to the scale of the project. However, issuance costs and project economics encourage grouping small projects into larger projects or programs of projects.

Timing and lifecycle: Although state law might authorize longer maturities, bonds are usually issued on a 20- to 30-year basis based on the economics of bond markets.

Ease of use: Issuing general obligation bonds requires specialized advisors. Fees and expenses associated with issuance, as well as capitalized interest during construction (if applicable), can be included in the principal of the debt.

---

[17] There are two types of municipal bonds: those that are general obligations of the issuer and those that are secured by specified revenue (revenue bonds, discussed in the following section).

[18] AASHTO Center for Excellence in Project Finance. "General Obligation Bonds." http://www.transportation-finance.org/funding_financing/financing/bonding_debt_instruments/municipal_public_bond_issues/general_obligation_bonds.aspx. Accessed August 24, 2012

[19] Ibid.

OTHER LIMITATIONS OF THE TOOL

General obligation bond transactions can be complex, requiring knowledge of leasing, real estate law, corporate entity formation, and securitization in addition to public finance and tax law. Therefore, for these instruments to be viable, the amount of debt to be issued needs to exceed its transaction costs.

USE OF THE TOOL IN PRACTICE

Example: Washington Dulles Corridor Metrorail Project

Location: Northern Virginia

Description: Voters in the Virginia counties of Arlington, Fairfax, Loudoun, and Prince William approved over $1.6 billion in general obligation bonds for transportation projects. About one-third of these bonds support a 23-mile Metrorail extension from Fairfax County to Loudoun County. General funds, which come from primarily local property taxes, are used to pay the debt service.[20]

GETTING STARTED

The process generally starts with a state, county, city, redevelopment agency, or special-purpose district identifying a project or a group of small projects. The agency will need to involve advisors, including a financial advisor, an underwriter, bond counsel, a rating agency, and insurers, to assess the feasibility of issuing a general obligation bond.

REFERENCES

AASHTO Center for Excellence in Project Finance. "General Obligation Bonds." http://www.transportation-finance.org/funding_financing/financing/bonding_debt_instruments/municipal_public_bond_issues/general_obligation_bonds.aspx. Accessed August 24, 2012

Biesiadny, Tom. "Presentation to Maryland Transit Funding Study Steering Committee." Fairfax County Department of Transportation. http://www.mdot.maryland.gov/Office%20of%20Planning%20and%20Capital%20Programming/Transit_Funding_Study/Documents/Fairfax%20County%20Presentation.pdf. Accessed August 23, 2012.

FHWA. *Innovative Finance Primer.* 2004. http://www.fhwa.dot.gov/ipd/pdfs/finance/ifprimer.pdf.

Title 26 *U.S. Code*, § 149. "Bonds must be registered to be tax exempt; other requirements." http://www.law.cornell.edu/uscode/html/uscode26/usc_sec_26_00000149----000-.html.

Good Jobs First. "Municipal Bond Basics: What are the Main Types of Municipal Bonds?" http://www.publicbonds.org/bond_basics/municipal_bonds.htm. Accessed August 22, 2011.

---

[20] Biesiadny, T. "Presentation to Maryland Transit Funding Study Steering Committee." Fairfax County Department of Transportation. http://www.mdot.maryland.gov/Office%20of%20Planning%20and%20Capital%20Programming/Transit_Funding_Study/Documents/Fairfax%20County%20Presentation.pdf. Accessed August 23, 2012.

B-3.   REVENUE BONDS

A revenue bond is a type of municipal bond that is secured by a specific revenue stream. Revenue bonds can be issued by cities, counties, and, in some states, special districts to finance improvements for a revenue-producing enterprise. Revenue bonds are repaid solely from the revenue generated by the financed facility (e.g., an airport, water system, or sewer system). The revenue used to back the bonds can include service charges or rates, tolls, connection fees, admission fees, and rents. Revenue bonds can finance transit facilities, with fare box revenue providing part of the revenue stream required to secure the bond.

Under the typical revenue bond structure, income from the revenue-generating enterprise is put into a revenue fund. Expenses for operations and maintenance are paid first from the revenue fund. Only after those costs are paid do revenue bondholders receive payments. Most project-backed revenue sources are less secure than taxes that would back a general obligation bond. In addition, revenue bonds are not backed by the full faith and credit of a public entity, as general obligation bonds are. For these reasons, revenue bonds carry a somewhat higher default risk and therefore higher interest rates than general obligation bonds.

KEY FACTORS FOR EVALUATION

Applicability to different types of infrastructure: Revenue bonds can be used only for revenue-generating infrastructure (e.g., parking, water and wastewater systems, toll roads and bridges, and transit).

Approval requirements and legal and political considerations: Some states require voter approval for revenue bonds.

Application for strong and weak real estate markets: This financing tool does not rely on new development and is therefore applicable in strong and weak real estate markets.

Capacity and scale: Revenue bonds can be sold in $5,000 units and have no explicit thresholds on the capacity and scale of a project. However, for smaller bond issuances, transaction costs can be considerable compared to total proceeds, so smaller projects are often grouped together for bond issuances.

Ease of use: Revenue bonds are typically part of a complex financing package that requires extensive financial analysis and bond counsel.

Timing and lifecycle: Revenue bonds typically mature in 20 to 30 years. However, all the bonds in an issuance might not mature at the same time. Bond issuances with staggered maturity dates are known as serial bonds.

OTHER LIMITATIONS OF THE TOOL

Revenue bonds are somewhat more risky than other types of municipal bonds because they are not backed by the full faith and credit of a public entity and therefore have higher interest rates.

USE OF THE TOOL IN PRACTICE

Example: West Dublin/Pleasanton BART Station and Transit Village

Location: Dublin, California

Description: Bay Area Rapid Transit (BART) recently completed the $106-million West Dublin/Pleasanton Station, a new transit station built on an existing commuter rail line. One of the most challenging aspects of the project was that structured parking needed to be built before other revenue-generating components of the project. BART used bond financing to build the station and structured parking. The debt service on the bonds will be repaid using proceeds from planned real estate development, as well as BART parking and fare box revenue. The cities of Dublin and Pleasanton and Alameda County agreed to place a total of $8 million in a reserve account, which will be used if there is a shortfall in the debt service on the bonds or in station operating costs.[21]

GETTING STARTED

Securing a revenue bond would require significant analysis of the project's legal, technical, and financial feasibility, including analysis to document project costs and projected revenue from the built infrastructure project.

REFERENCES

BART. "BART riders celebrate grand opening of West Dublin/Pleasanton Station Saturday." February 18, 2011. http://www.bart.gov/news/articles/2011/news20110218.aspx.

Center for Transit-Oriented Development. *Capturing the Value of Transit*. Prepared for FTA. 2008. http://reconnectingamerica.org/assets/Uploads/ctodvalcapture110508v2.pdf.

---

[21] BART. "BART riders celebrate grand opening of West Dublin/Pleasanton Station Saturday." February 18, 2011. http://www.bart.gov/news/articles/2011/news20110218.aspx.

B-4. PRIVATE ACTIVITY BONDS

Private activity bonds (PABs) are federal- and state-tax-exempt securities issued by state or municipal governments to provide financing for private entities.[22] The federal government imposes a limit on how many PABs each state can issue annually based on the state's population.[23] Although programs vary by state, PABs are used to finance projects with a public benefit such as low-income housing development, hazardous and solid waste facilities, redevelopment projects, and infrastructure projects like sewer, water, and energy systems. PABs are secured by and repaid from revenue generated by the project that they financed. PABs are not backed or guaranteed by the issuing municipality.

PABs were not available to finance transportation infrastructure until the 2005 Safe, Accountable, Flexible, Efficient Transportation Equity Act: A Legacy for Users (SAFETEA-LU) added transportation infrastructure to the types of projects that can be funded with PABs and set aside $15 billion for DOT to allocate among qualified projects. This $15 billion is not subject to any individual state's PAB volume cap. By providing low-cost financing to projects with private involvement, the DOT PAB program aims to increase private investment in transportation infrastructure. As of May 2011, 30 percent of the $15 billion had been allocated to seven projects, leaving just over $10 billion of PABs for future projects.[24]

This tool description focuses on DOT's PAB program; local governments should check with their state bond allocation offices for information on state programs.

KEY FACTORS FOR EVALUATION

Applicability to different types of infrastructure: To be eligible for a PAB, projects must receive federal assistance under Title 23 (Highways) or 49 (Transportation) of the United States Code (U.S.C.) and involve private participation.[25] PABs are thus a form of public-private partnership. Qualified projects under Titles 23 and 49 include construction of bicycle transportation and pedestrian walkways along urban and rural principal arterial routes, acquisition of scenic easements and scenic or historic sites, landscaping and other scenic beautification, preservation of abandoned railway corridors (including the conversion into pedestrian or bicycle trails), environmental mitigation to address water pollution due to highway runoff, intra- or intercity bus terminals, bus corridors, and parking facilities.

DOT has the discretion to determine what is considered federal assistance for eligibility purposes. In general, some material element of a project, such as engineering design work, must be supported by federal funds. Transportation Infrastructure Finance and Innovation Act (TIFIA)[26] assistance is considered federal assistance for purposes of PAB eligibility; both TIFIA and PABs can be used in the same project.

---

[22] FHWA. "Private Activity Bonds (PABs)." http://www.fhwa.dot.gov/ipd/fact_sheets/pabs.htm. Accessed August 22, 2011.

[23] Commonwealth of Massachusetts. "Private Activity Bonds." http://www.mass.gov/anf/budget-taxes-and-procurement/cap-finance/private-activity-bonds.html. Accessed August 23, 2012.

[24] FHWA. "Private Activity Bonds (PABs)." http://www.fhwa.dot.gov/ipd/fact_sheets/pabs.htm. Accessed August 22, 2011.

[25] The private sector must use at least 10 percent of the financing and pay at least 10 percent of the debt.

[26] See Section C-2 of this appendix for a description of TIFIA.

Approval requirements and legal and political considerations: DOT allocates the $15 billion of PAB assistance authorized in 2005. However, PAB allocation from DOT merely provides a "license to issue." Upon receipt of PAB allocation, the private entity must still identify the appropriate public-sector issuer for a PAB and follow all requirements for the bond issuances.

Application for strong and weak real estate markets: PAB issuance can be affected by the conditions of the real estate market if a share of the revenue stream to serve the debt is real estate-related, such as real estate taxes. In a weak real estate market, investors might require the issuer to use credit assistance to prevent default.

Capacity and scale: PABs have no explicit eligibility requirements related to the capacity and scale of a project. However, most of the PAB allocations have gone to large projects. The smallest PAB allocation to a single project was $398 million; the median allocation amount is $592 million.

Timing and lifecycle: The term to maturity cannot exceed 120 percent of the useful life of the facility being financed. As a part of the application to DOT, the applicant needs to identify the financial structure of the proposed project, including the sources of security and repayment for the bonds. Repayment of the bonds is typically tied to the revenue generated by the project.

Ease of use: PABs require that projects to be financed involve private participation in addition to receiving federal assistance under U.S.C. Titles 23 or 49. As such, project elements funded with federal assistance must follow all federal-aid requirements. In addition, PAB allocation recipients are required to retain bond counsel to ensure that all Internal Revenue Service requirements for PABs are followed. Financing with PABs requires strict compliance with requirements and limitations established by the Internal Revenue Code.

OTHER LIMITATIONS OF THE TOOL

While PABs are intended to lower financing costs by offering tax-exempt benefits, depending on market demand and conditions, PABs can be more expensive financing tools than traditional tax-exempt bonds or other alternatives because their interest rate advantage can be achieved only if the PAB receives an investment grade rating from one of the nationally recognized rating agencies.

USE OF THE TOOL IN PRACTICE

Between the passage of SAFETEA-LU in 2005 and June 2012, PABs have been issued for six projects, and DOT has approved allocations for an additional seven. Together these thirteen projects account for just over half of the $15 billion in PABs allowed under the act (shown in Exhibit B-2).

Project	PAB Allocation
Bonds Issued	
Capital Beltway HOT Lanes, VA	$ 589 million
North Tarrant Express, TX	$ 400 million
IH 635 (LBJ Freeway), TX	$ 615 million
Denver RTD Eagle Project, CO	$ 398 million
CenterPoint Intermodal Center, Joliet, IL	$ 150 million
Downtown Tunnel/Midtown Tunnel, Norfolk, VA	$ 664 million
Subtotal	**$ 2,816 million**
Allocations	
Knik Arm Crossing, AK	$ 600 million
CenterPoint Freight Intermodal Center, Joliet, IL	$ 1,190 million
I-80 RailPort, Seneca, IL	$ 576 million
CenterPoint Intermodal Center, Kansas City, KS	$ 475 million
Northwest Corridor Project, GA	$700 million
I-95 HOV/HOT project	$600 million
RidgePort Logistics Center, Will County, IL	$ 555 million
Subtotal	$ 4,696 million
Total PAB Allocations and Issues	**$ 7,511 million**

Exhibit B-2. PAB Pipeline as of June 6, 2012
Source: FHWA. "Tools & Programs: Federal Debt Financing Tools." http://www.fhwa.dot.gov/ipd/finance/tools_programs/federal_debt_financing/private_activity_bonds/index.htm. Accessed July 17, 2012.

Example: Denver Eagle Public-Private Partnership

Location: Denver, Colorado

Description: The population of the Denver metropolitan area is projected to grow 154 percent between 2000 and 2020. After considering options to expand the existing infrastructure to serve the future population, Denver's Regional Transportation District (RTD) decided to build 122 miles of new commuter and light rail, 18 miles of bus rapid transit service, and 21,000 new parking spaces at bus and rail stations.[27]

Due to the size of the expansion, RTD decided to finance and build the project in stages. The Eagle segment, consisting of the 22.8-mile East Corridor and the 11.2-mile Gold Corridor, will be built under a public-private partnership. The private sector is responsible for designing, building, partially financing, and operating the system, but RTD will retain ownership of all assets.[28] The concession term (i.e., the time the private sector will be responsible for the project) is 34 years—five years for construction and 29 years for operation. The total financing package is $1.64 billion. The private sector is contributing $54.3 million in equity. RTD issued $397.8 million in private activity bonds and is providing $1.14 billion in construction payments and $44 million in pre-completion service payments. In addition, RTD will make

---

[27] Metropolitan Planning Council. "PPP Profiles: EagleP3, a section of Denver's comprehensive transit expansion, FasTracks." http://www.metroplanning.org/news-events/article/6139. Accessed July 17, 2012.

[28] RTD. "Eagle P3 Project" http://www.rtd-fastracks.com/main_126. Accessed August 22, 2011.

monthly service payments to the developers after completion based on performance. The private-sector partners will have to comply with specific standards (e.g., response times to incidents and infrastructure maintenance). When the private partner does not meet those standards, RTD can reduce its service payments. RTD expects to pay a maximum of $5.5 billion in service payments over the life of the contract.[29]

GETTING STARTED

PABs can be issued only for projects with private investment and a stable revenue stream (e.g., user fees such as tolls or fares or other committed public sources). The first steps to request PAB assistance include:

- Assessment of project eligibility. Projects must receive federal assistance under U.S.C. Titles 23 or 49.

- Application submission. DOT accepts applications for PAB assistance from public agencies throughout the year. There is no fixed format for PAB applications; however, DOT has suggested that project sponsors provide the following information: amount of allocation requested, proposed date of bond issuance, draft bond counsel opinion letter, project description, description of Title 23 or 49 funding received by the project, and project readiness.[30]

REFERENCES

Commonwealth of Massachusetts. "Private Activity Bonds." http://www.mass.gov/anf/budget-taxes-and-procurement/cap-finance/private-activity-bonds.html. Accessed August 23, 2012.

FHWA. "Private Activity Bonds (PABs)." http://www.fhwa.dot.gov/ipd/fact_sheets/pabs.htm. Accessed August 22, 2011.

FHWA. "Tools and Programs: Private Activity Bonds (PABs)." http://www.fhwa.dot.gov/ipd/p3/tools_programs/pabs.htm. Accessed July 17, 2012.

FHWA. "Tools & Programs: Federal Debt Financing Tools." http://www.fhwa.dot.gov/ipd/finance/tools_programs/federal_debt_financing/private_activity_bonds/index.htm. Accessed July 17, 2012.

Metropolitan Planning Council. "PPP Profiles: EagleP3, a section of Denver's comprehensive transit expansion, FasTracks." http://www.metroplanning.org/news-events/article/6139. Accessed July 17, 2012.

RTD. "Eagle P3 Project" http://www.rtd-fastracks.com/main_126. Accessed August 22, 2011.

---

[29] Metropolitan Planning Council. "PPP Profiles: EagleP3, a section of Denver's comprehensive transit expansion, FasTracks." http://www.metroplanning.org/news-events/article/6139. Accessed July 17, 2012.

[30] FHWA. "Tools and Programs: Private Activity Bonds (PABs)." http://www.fhwa.dot.gov/ipd/p3/tools_programs/pabs.htm. Accessed July 17, 2012.

B-5.   CERTIFICATES OF PARTICIPATION AND LEASE REVENUE BONDS

Certificates of participation (COPs) and lease revenue bonds (LRBs) are tax-exempt bonds usually secured with revenue from an equipment or facility lease.[31] These instruments are issued by state-authorized entities (e.g., state public works boards, joint powers authorities, municipalities, or transit agencies).

KEY FACTORS FOR EVALUATION

Applicability to different types of infrastructure: LRBs and COPs have been used in public finance to support a broad variety of projects and programs, including acquisition of land or equipment, transportation (e.g., light rail and toll bridges), water and wastewater treatment facilities, and real estate (e.g., parking facilities, public buildings). Neither tool is suitable for funding operations and maintenance activities.

Approval requirements and legal and political considerations: LRBs and COPs do not require voter approval because these instruments are not backed by federal or state government revenue.

Application for strong and weak real estate markets: The issuance of COPs and LRBs can be affected by the conditions of the real estate market if the revenue stream to serve the debt is real estate-related, such as the lease of a public building with commercial space. In a weak real estate market, investors might require credit assistance to prevent default, including rental interruption insurance and the creation of a debt service reserve fund, which would be funded by the bond issue and is considered a last resort to pay debt service.[32]

Capacity and scale: There is no limit to the scale of projects; however, issuance costs and project economics encourage grouping small projects into larger projects or programs of projects to be funded. Local agencies with projects that are too small to attract investors or to otherwise be feasible for lease financing can work together and create a pool of projects to issue a COP or LRB. By grouping the projects, local agencies can minimize the issuing costs and possibly reduce the interest that must be paid on the lease. Additionally, because using COPs or LRBs allows the project to be financed by many small investors rather than one large one, it increases the pool of potential investors.

Timing and lifecycle: The maturity of the instrument cannot exceed the useful life of the facility or equipment funded. For instance, COP transactions involving FTA grants have funded multiple bus acquisitions with maturities up to 12 years and a light rail project with a maturity of 27.5 years.

Ease of use: The issue of both instruments requires participation of specialized advisors. Fees and expenses associated with issuance, as well as capitalized interest during construction (if applicable) can be included in the principal of the debt issued.

---

[31] AASHTO Center for Excellence in Project Finance. "Certificates of Participation." http://www.transportation-finance.org/funding_financing/financing/bonding_debt_instruments/certificates_of_participation.aspx. Accessed August 24, 2012.

[32] Tax law sets the amount of the reserve fund. For state lease financings, it is generally 50 percent of the maximum semi-annual debt service.

OTHER LIMITATIONS OF THE TOOL

COP and LRB transactions can be complex, requiring knowledge of leasing, real estate law, corporate entity formation, and securitization, in addition to public finance and tax law. Therefore, their costs of issuance tend to be high and are worthwhile only for larger projects or groups of smaller projects.

USE OF THE TOOL IN PRACTICE

Example: Sacramento Regional Transit District's light-rail system

Location: Sacramento, California

Description: In 1985, the city of Sacramento issued $29.4 million of COPs for funding needed to complete the Sacramento Regional Transit District's light-rail system when costs rose above the original project estimate of $131.2 million (Exhibit B-3). The city's share of the project rose from 5.1 percent to 19 percent due to the cost overruns.[33]

Exhibit B-3. Sacramento Regional Transit District light rail.
Source: Ian Britton via FreeFoto.com.

Example: Metrorail Garage Projects

Location: Montgomery County, Maryland

Description: In 2011, Montgomery County issued around $35 million in LRBs to finance the costs of parking structures at transit stations and to refund outstanding LRBs to save money on their debt service. The parking structures will serve stations for Metrorail, the Washington, D.C., regional rail system. The county expects to pay debt service on the LRBs from rental payments made by the Washington Metropolitan Area Transit Authority (WMATA) under a lease agreement. Parking revenue is deposited into an account established for the county and held by WMATA in trust to meet the transit authority's obligation under the facility lease agreements.[34]

GETTING STARTED

LRBs and COPs can be issued by multiple entities, including joint powers authorities, municipalities, transit agencies, or counties, depending on state law. The entity identifies a project or a group of small projects suitable for a lease arrangement. In addition, the issuing entity will need a group of advisors, including a financial advisor, an underwriter, bond counsel, a rating agency, and insurers, to assess the feasibility of issuing either a LRB or a COP.

---

[33] FTA. "FY 2013 Funding Recommendations." http://www.fta.dot.gov/printer_friendly/12868_3535.html. Accessed August 24, 2012.

[34] Montgomery County, Maryland. *Lease Revenue Project and Refunding Bonds (Metrorail Garage Projects).* 2011.

REFERENCES

AASHTO Center for Excellence in Project Finance. "Certificates of Participation." http://www.transportation-finance.org/funding_financing/financing/bonding_debt_instruments/certificates_of_participation.aspx. Accessed July 25, 2011.

FTA. "FY 2013 Funding Recommendations." http://www.fta.dot.gov/printer_friendly/12868_3535.html. Accessed August 24, 2012.

Montgomery County, Maryland. *Lease Revenue Project and Refunding Bonds (Metrorail Garage Projects)*. 2011. http://bonds.montgomerycountymd.gov/data/MDMontgomery04a-FIN.pdf.

State of California. "Lease Revenue Bonds." http://sam.dgs.ca.gov/TOC/6000/6872.htm. Accessed July 25, 2011.

B-6.   REVOLVING LOAN FUNDS

A revolving loan fund (RLF) is a pool of money dedicated to specific kinds of investments. The money used to repay loans replenishes the fund and is loaned out again.[35] RLF initial funding sources are typically public or private "seed money" and/or an ongoing revenue stream. The capitalization or initial funding could come from appropriations, grants, borrowing of capital funds, or the proceeds of a one-time asset sale. The ongoing revenue stream could be a dedicated portion of an existing or new tax.

Government agencies or nonprofits establish RLFs to help projects move forward by providing access to capital funds through a variety of financing mechanisms, including loans with rates and repayment terms that could be more favorable than the borrower could find in the market and credit assistance tools such as letters of credit,[36] lines of credit,[37] bond insurance, debt service reserves, and debt service guarantees. RLFs can provide access to capital markets for projects that have poor risk profiles to meet economic development (e.g., new business development), environmental (e.g., safe drinking water), or other public policy goals. RLF financing can also be useful for projects where the revenue stream might be irregular. RLF customers can include local governments, special districts, state agencies, private corporations, or nonprofit organizations.

KEY FACTORS FOR EVALUATION

Applicability to different types of infrastructure: The types of projects funded depend on the RLF's policies. RLFs have been established to fund affordable housing, historic preservation, transportation, energy efficiency, safe drinking water, and small business development.[38] RLFs have been used to fund infrastructure projects in the water, transportation, environmental, and agricultural sectors.

Approval requirements and legal and political considerations: The need for legislation to establish an RLF varies by state. In addition other barriers might exist in current law, regulation, or policy related to bonding, making loans, or other financial assistance.

Application for strong and weak real estate markets: Real estate market strength does not directly affect an RLF unless the debt repayment source is related to property or real estate (e.g., property taxes). If the real estate market is weak, the borrower's default risk increases, and the fund's long-term sustainability can be undermined.

Capacity and scale: The scale of financed projects varies depending on the fund's policies. Loans can range from $100,000 to tens of millions of dollars.

---

[35] EPA. *A Sustainable Brownfields Model Framework*. 1999. http://www.epa.gov/nscep/index.html. p. 121.

[36] A letter of credit is a form of loan from the SIB to be used only in the event of a shortfall in net revenue for debt service (i.e., a contingent loan). A letter of credit is security provided directly to the lender and/or bondholders (via the trustee) rather than to the borrower or project sponsor.

[37] A line of credit is a contingent loan similar to a letter of credit. The difference is that a line of credit is security available directly to the borrower or project sponsor with flexibility in use of the funds.

[38] Booth, S.; Doris, E.; Knutson, D.; and Regenthal, S. *Using Revolving Loan Funds to Finance Energy Savings Performance Contracts in State and Local Agency Applications*. National Renewable Energy Laboratory. 2011.

Timing and lifecycle: For infrastructure projects, the loan term can range between 10 and 30 years.

Ease of use: RLFs can be an effective tool to help infrastructure projects advance. However, creating an RLF requires consensus on numerous institutional, financial, and managerial decisions that can involve several stakeholders such state agencies, private donors, and potential users. In addition, how the RLF is managed largely determines the fund's long-term sustainability. While it might be tempting to leverage an RLF as much as possible, higher leverage also leads to higher risk exposure and can make the fund less sustainable in the long run.

OTHER LIMITATIONS OF THE TOOL

A major challenge for RLFs is to provide access to capital funds to projects that could create social benefits but have a poor risk profile due to a lower likelihood of loan repayment (typically non-revenue generating projects). To be self-replenishing, RLFs must generate enough return in interest and principal payments. Therefore, an RLF's success depends largely on enforcing loan repayments; any defaults on project loans will reduce the number of additional loans that can be made. The RLF's ability to sustain itself relies heavily on the fund managers' ability to mitigate the risk of poor risk profile projects. Strategies could include investing in diverse sectors or projects and requiring additional credit assistance.

USE OF THE TOOL IN PRACTICE

No RLFs focus specifically on TOD infrastructure, but because many state infrastructure banks are structured as RLFs, any bank that finances transit facilities, bikeways, streets renovation, and similar projects encourages TOD. The Minnesota Transportation Revolving Loan Fund, a state infrastructure bank described in the next tool profile, is one example.

GETTING STARTED

Implementing a RLF requires making multiple institutional, financial, and managerial decisions:[39]

- **RLF policy:** Stakeholders need to decide the purpose of the fund, the types of projects to fund, and the type of clients that will have access to capital.

- **Enabling legislation:** Stakeholders need to determine whether legislation is needed and assess any potential barriers in current law, regulation, or policy related to bonding, making loans, or other financial assistance.

- **Capitalization:** The entity establishing the fund needs to identify initial capitalization sources (seed money). Another option is to establish an ongoing revenue stream to support the fund's activities. Factors to consider when capitalizing a RLF include the estimated cost of identified first-round projects; the estimated cost of potential future projects; forms of assistance to be provided (loans and terms, lines of credit, guarantees, etc.). Additionally, the ability and suitability to leverage the

---

[39] FHWA. "Resources: Federal Credit Assistance Tools." http://www.fhwa.dot.gov/ipd/finance/resources/federal_credit/sib_primer.htm#IV2a. Accessed July 25, 2011.

bank's capital should be discussed. Bond issuance has been one of the mechanisms used to increase the pool of resources available for projects, but this is subject to the issuer's credit rating (bonds are required to be investment grade, at least BBB) and current market conditions.

- **Institutional structure:** An institution will have to be selected to manage the fund. Many options are viable, including establishing a single purpose finance authority for the RLF or housing the fund's account in a state agency such as the state treasury department, economic development agency, or department of commerce.

- **Project screening criteria:** Projects initially are screened based on general eligibility guidelines, which might include: the existence and strength of identified revenue streams for loan or other fund assistance repayment, consistency with local and regional plans, and consistency with state infrastructure plans.

- **Project selection criteria:** Following project screening and capitalization decisions, eligible projects are subject to a more detailed project selection procedure based on specific criteria. Potential criteria could include: the project's economic and social benefits, the project's impact on public mobility and safety, the project's readiness (completion of environmental clearance and construction approvals), and the project's ability to leverage other funding sources.

- **Loan application material:** Fund managers will need to determine what supporting documents and materials will be required with the loan application.

REFERENCES

Booth, S.; Doris, E.; Knutson, D.; and Regenthal, S. *Using Revolving Loan Funds to Finance Energy Savings Performance Contracts in State and Local Agency Applications*. National Renewable Energy Laboratory. 2011. http://www.nrel.gov/applying_technologies/state_local_activities/pdfs/51399.pdf.

EPA. *A Sustainable Brownfields Model Framework*. 1999. http://www.epa.gov/nscep/index.html.

FHWA. "Resources: Federal Credit Assistance Tools." http://www.fhwa.dot.gov/ipd/finance/resources/federal_credit/sib_primer.htm#IV2a. Accessed July 25, 2011.

Minnesota Department of Transportation. "Transportation Revolving Loan Fund—How it works." http://www.dot.state.mn.us/planning/program/trlf_how.html. Accessed July 25, 2011.

B-7.   STATE INFRASTRUCTURE BANKS

State infrastructure banks (SIBs) help finance infrastructure projects through a variety of mechanisms, including loans with rates and repayment terms that are better than the borrower could find in the market and credit assistance tools such as letters of credit, lines of credit, bond insurance, debt service reserves, and debt service guarantees. The money used to repay loans goes to the bank for additional lending. SIBs thus function as a type of RLF. SIB customers include local governments, special districts, state agencies, private corporations, and nonprofit organizations.

SIBs were first authorized in 1995 as a part of the National Highway Designation Act to help accelerate transportation improvements by offering various forms of financial assistance to local entities through state transportation departments. During the pilot program, 10 states were permitted to use a portion of their federal-aid funds, matched with non-federal funds, as seed money to establish an SIB. [40] SAFETEA-LU authorized all states to enter into cooperative agreements with DOT to establish infrastructure revolving loan funds eligible to be capitalized with federal transportation funds authorized for fiscal years 2005 to 2009.

As of 2011, 33 state departments of transportation, including those in Colorado, Utah, and Georgia, have used federal funding to create SIBs. The amount of funding available depends on the size of the state's SIB, which varies from less than $1 million (Wisconsin) to more than $100 million (Ohio). [41] To leverage the bank's lending capacity, some states have issued debt through bonds.

KEY FACTORS FOR EVALUATION

Applicability to different types of infrastructure: Projects must be eligible under U.S.C. Title 23 or Title 49. Eligible transit projects include acquisition of real estate property and rights-of-way; capital projects to modernize existing fixed guideway systems; capital projects to replace, rehabilitate, and purchase buses and related equipment and to construct bus-related facilities; construction of dedicated bus and high occupancy vehicle lanes; and construction of park and ride lots.

The types of projects funded vary according to the SIB's loan policies, which depend on each state's needs and strategies. Some banks, such as those in South Carolina and Florida, seek to leverage the available capital as much as possible and offer loans only to projects that have strong supplementary financing sources and proven revenue streams. Borrowers can use several revenue sources, such as tolls, user fees, or other taxes, to repay loans. Other states, such as Minnesota, do not lend to revenue-generating projects such as toll roads but and provide financing only to public agencies that can make repayments using future local government revenue.

---

[40] FHWA. "Tools & Programs: Federal Credit Assistance Tools. *State Infrastructure Banks (SIBs).*" http://www.fhwa.dot.gov/ipd/finance/tools_programs/federal_credit_assistance/sibs/index.htm. Accessed August 20, 2012.

[41] West Coast Collaborative. "State Infrastructure Banks (SIBs) Fact Sheet." http://westcoastcollaborative.org/files/meetings/2006-02-01/SIBs%20Fact%20Sheet.pdf. Accessed August 20, 2012.

Other states have established specific programs to help regional transportation projects that meet certain conditions. For example, Florida set aside $100 million for regional projects on the condition that at least 25 percent of the costs be matched by other sources. In addition, Florida's SIB provides emergency loans for disaster damage to transportation infrastructure.

Approval requirements and legal and political considerations: State infrastructure banks will generally lend to only certain types of entities. In addition, users will have to meet the specific eligibility requirements and underwriting criteria of the infrastructure bank.

Application for strong and weak real estate markets: This tool applies in both strong and weak real estate markets unless the debt repayment source is related to property or real estate (e.g., property taxes). Other sources of debt repayment could include special assessments, future federal or state funds, and user fees from revenue-generating projects.

Capacity and scale: The scale of financed projects varies depending on the fund's policies. Loans can range from $100,000 to tens of millions of dollars.

Timing and lifecycle: The maximum loan term varies from 10 years (e.g., Ohio) to 30 years (e.g., Florida), and the loan rate may be set at or below market rate. Some states allow payments to be deferred up to 5 years.

Ease of use: The ability to borrow from an SIB will be determined by the SIB's loan policies and procedures. Some states have established accounts in their SIBs that are funded only with state funds to have more flexible loan policies and avoid federal restrictions applicable to the federally funded portion.

OTHER LIMITATIONS OF THE TOOL

A major challenge for SIBs is to provide access to capital funds to projects that could create social benefits but have a poor risk profile due to a lower likelihood of loan repayment. Non-revenue generating projects need to be supported by alternative sources that are not necessarily reliable unless government entities commit to repaying the loan.

USE OF THE TOOL IN PRACTICE

Example: South Carolina SIB

Location: South Carolina

Description: The South Carolina Transportation Infrastructure Bank issued a series of bonds in amounts ranging from $270 million to $370 million between 1998 and 2003 that were supported primarily by state truck registration fees, loan repayments from the counties, federal highway program apportionments, and non-tax revenue funding from the South Carolina Department of Transportation.

Example: Minnesota SIB

Location: Minnesota

Description: Minnesota's SIB, known as the Transportation Revolving Loan Fund, was established in 1997. The legislation authorized the Minnesota Department of Transportation (MnDOT), the Minnesota Department of Trade and Economic Development, and the Minnesota Public Facilities Authority to jointly develop and administer a SIB program. In June 1997, the federal government authorized Minnesota to create a SIB program and appropriated $3.96 million to capitalize the fund with the requirement of a non-federal match of 25 percent of the federal contribution.[42]

The fund provides loans, loan guarantees,[43] lines of credit, credit enhancements, equipment financing leases, bond insurance, and other forms of financial assistance. Cities, counties, and other governmental entities can borrow from the fund. Private entities are not themselves directly eligible for financing, but they could enter into agreements with eligible borrowers to finance projects. Eligible projects include pre-design studies; acquisition of right-of-way; rail safety projects; and transit capital purchases and leases. Borrowers must secure their loans by providing a local general obligation bond as collateral. Possible loan repayment sources include special assessments, property tax levies, TIF, local government option sales taxes, future federal funds, future state funds, and customer fees from revenue-generating projects.[44]

GETTING STARTED

Application procedures will vary by state.

REFERENCES

FHWA. "Resources: Federal Credit Assistance Tools." http://www.fhwa.dot.gov/ipd/finance/resources/federal_credit/sib_primer.htm#IV2a. Accessed July 25, 2011.

FHWA. "Tools & Programs: Federal Credit Assistance Tools. *State Infrastructure Banks (SIBs)*." http://www.fhwa.dot.gov/ipd/finance/tools_programs/federal_credit_assistance/sibs/index.htm. Accessed August 20, 2012.

Minnesota Department of Transportation. "Transportation Revolving Loan Fund—How it works." http://www.dot.state.mn.us/planning/program/trlf_how.html. Accessed July 25, 2011.

West Coast Collaborative. "State Infrastructure Banks (SIBs) Fact Sheet." http://westcoastcollaborative.org/files/meetings/2006-02-01/SIBs%20Fact%20Sheet.pdf. Accessed August 20, 2012.

---

[42] Minnesota Department of Transportation. "Transportation Revolving Loan Fund—How it works." http://www.dot.state.mn.us/planning/program/trlf_how.html. Accessed July 25, 2011.

[43] Loan guarantees are contract(s) entered into by the SIB in which the SIB agrees to take responsibility for all or a portion of a project sponsor's financial obligations for a project under specified conditions.

[44] Minnesota Department of Transportation. "Transportation Revolving Loan Fund—How it works." http://www.dot.state.mn.us/planning/program/trlf_how.html. Accessed July 25, 2011.

B-8.   GRANT ANTICIPATION REVENUE VEHICLE BONDS

Grant anticipation revenue vehicle bonds (GARVEEs)[45] are federal-tax-exempt debt mechanisms[46] (e.g., bonds, notes, certificates, mortgages, or leases) that are backed by future Title 23 federal transportation funding.[47]

This financing mechanism is suitable when a state cannot construct projects using traditional pay-as-you-go funding.[48] GARVEE financing allows the state to use future federal highway funds as the revenue stream to pay debt service. This tool allows the state to accelerate construction timelines and spread the cost of a transportation facility over its useful life. GARVEEs expand access to capital markets as an alternative or in addition to potential general obligation or revenue bonding capabilities. The benefit of upfront monetization of federal funds needs to be weighed against the cost of consuming a portion of future appropriations to pay debt service.[49]

A state, political subdivision, or public authority such as an SIB can issue GARVEEs. No federal prohibition or restriction prevents a local government from issuing a GARVEE. However, local governments might face more legal and financial requirements from investors than state governments since they are perceived as higher risk than state governments.

KEY FACTORS FOR EVALUATION

Applicability to different types of infrastructure: GARVEE debt financing can fund projects (or programs of projects) eligible under U.S.C. Title 23, which include bicycle transportation infrastructure and pedestrian walkways, beautification of streets, construction of publicly owned intra- or intercity bus terminals, and environmental mitigation to address water pollution. Parking facilities are subject to DOT approval.

Approval requirements and legal and political considerations: The Federal Highway Administration (FHWA) determines the eligibility of projects to be financed with GARVEEs. States approve the bond issue, which might require enabling legislation. In general, GARVEE debt can be issued without voter approval; however, some states require report or approval of administrative bodies when issuing GARVEE debt. Additionally, some states have established caps on the volume of GARVEE debt that can

---

[45] U.S.C. § 122 of Title 23 provides the federal legislative framework for this tool.

[46] In many cases, they are exempt from state tax as well.

[47] California Department of Transportation. "Memorandum: Approval of Documents Related to the Bond Issuance of Grant Anticipated Revenue Vehicle Bonds, Series 2008A Resolution FG-08-01." August 27-28, 2008.

[48] AASHTO Center for Excellence in Project Finance. "Grant Anticipation Revenue Vehicles." http://www.transportation-finance.org/funding_financing/financing/bonding_debt_instruments/municipal_public_bond_issues/garvees.aspx. Accessed July 25, 2011.

[49] Mercator Advisors LLC. "Evaluation of Innovative Finance Tools as a Transportation Financing Mechanism." Commission Briefing Paper 5A-13. January 10, 2007.

be issued. For instance, in the state of Louisiana, the amount of resources available for enhancement projects is capped at 10 percent of future proceeds from DOT.[50]

Application for strong and weak real estate markets: A key factor that affects the risk profile of this instrument is the structure of the revenue pledge since the federal government is not guaranteed to provide the expected financing to pay GARVEE debt. To mitigate this risk, states can pledge additional revenue sources as a back-up. For instance, they can pledge state fuel tax revenue or local property taxes as a secondary source of revenue to pay for the debt service if future federal-aid highway funds are not available. Generally, secondary sources of revenue can result in lower interest costs on the bonds, as the market might perceive less risk of default. In a weak real estate market, investors could perceive these secondary sources as uncertain, and they might require additional credit assistance such as reserves or higher debt service to coverage ratio to prevent default.

Capacity and scale: Common characteristics for GARVEE financing projects are:

- They are large enough to benefit from borrowing rather than traditional pay-as-you-go funding.

- The costs of delaying project construction offset the costs of GARVEE financing.

- They do not have access to a revenue stream (e.g., local taxes or tolls), and other forms of repayment (e.g., state appropriations) are not feasible.

- The sponsors (generally state departments of transportation) agree to set aside future federal-aid highway funds to satisfy debt service requirements.

Timing and lifecycle: GARVEE debt terms are very flexible. Generally, the issuer and the capital markets can determine the basic metrics such as debt service coverage ratios, interest rates, maturity, debt reserves, or insurances. In other cases, state legislation sets the maximum term available. For instance, California Transportation Commission GARVEE bonds have a 12-year maturity.[51]

Ease of use: The issue of GARVEE debt requires the participation of specialized agents; associated fees and expenses can be reimbursed by the federal government.

USE OF THE TOOL IN PRACTICE

Example: Fast Forward - Georgia's Congestion Relief Program

Main Agency: Georgia Department of Transportation

---

[50] AASHTO Center for Excellence in Project Finance. "GARVEE Bonds." http://www.transportation-finance.org/funding_financing/legislation_regulations/state_local_legislation/enabling_legislation_federal_programs/garvee.aspx. Accessed July 25, 2011.

[51] Lockyer, B. "Analyses of GARVEE Bonding Capacity 2012." California State Treasurer. 2012. p.5.

Description: Georgia's Fast Forward, a six-year, $15.5-billion transportation program introduced in 2004, aims to add capacity to Georgia's highways and improve the existing highway network so that it operates more efficiently.[52]

Projects selected for the program are intended to provide short- and long-term congestion relief. Among the projects is a $286-million bus rapid transit project on two heavily congested corridors in the Atlanta area. Fast Forward is funded using GARVEE bonds, general obligation bonds, guaranteed revenue bonds, and federal funds. The agencies involved in Fast Forward are the Atlanta Regional Commission, the Georgia State Road and Tollway Authority, the Georgia Regional Transportation Authority, and local governments.[53]

GETTING STARTED

State, political subdivisions of a state, and public authorities are eligible issuers of GARVEEs. State law can authorize other entities. Public entities who would like to have access to GARVEE financing mechanisms need to determine:

- Project eligibility; the project or projects must qualify under U.S.C. Title 23 and be large enough to benefit from borrowing rather than traditional pay-as-you-go funding.

- The feasibility of accessing future federal-aid highway funds.

- The legal framework for issuing GARVEE debt, which is generally regulated by state law. Some states require enabling legislation to issue GARVEE debt, while others have established caps on the volume of GARVEE debt that can be issued.

REFERENCES

AASHTO Center for Excellence in Project Finance. "GARVEE Bonds." http://www.transportation-finance.org/funding_financing/legislation_regulations/state_local_legislation/enabling_legislation_federal_programs/garvee.aspx. Accessed July 25, 2011.

AASHTO Center for Excellence in Project Finance. "Grant Anticipation Revenue Vehicles." http://www.transportation-finance.org/funding_financing/financing/bonding_debt_instruments/municipal_public_bond_issues/garvees.aspx. Accessed July 25, 2011.

California Department of Transportation. "Memorandum: Approval of Documents Related to the Bond Issuance of Grant Anticipated Revenue Vehicle Bonds, Series 2008A Resolution FG-08-01." August 27-28, 2008. http://www.dot.ca.gov/hq/transprog/ctcbooks/2008/0808/12_4.6.pdf.

FHWA. *Innovative Finance Primer*. 2004. http://www.fhwa.dot.gov/ipd/pdfs/finance/ifprimer.pdf.

---

[52] Georgia Department of Transportation. "FAST Forward - Georgia's Congestion Relief Program." http://www.dot.state.ga.us/informationcenter/programs/transportation/fastforward/Pages/default.aspx. Accessed July 25, 2011.

[53] Ibid.

FHWA Resource Center. *What Every Transportation Manager Should Know about GARVEEs*. 2007. http://www.fhwa.dot.gov/ipd/pdfs/finance/Federal_Debt_Financing_Tools_120109.pdf.

Georgia Department of Transportation. "FAST Forward - Georgia's Congestion Relief Program." http://www.dot.state.ga.us/informationcenter/programs/transportation/fastforward/Pages/default.aspx. Accessed July 25, 2011.

Lockyer, Bill. "Analyses of GARVEE Bonding Capacity 2012." California State Treasurer. 2012. http://www.treasurer.ca.gov/bonds/garvee.pdf.

Mercator Advisors LLC. "Evaluation of Innovative Finance Tools as a Transportation Financing Mechanism." Commission Briefing Paper 5A-13. January 10, 2007. http://transportationfortomorrow.com/final_report/pdf/volume_3/technical_issue_papers/paper5a_13.pdf.

Title 23 *U.S.Code*, § 122. "Payments to States for bond and other debt instrument financing." http://www.law.cornell.edu/uscode/text/23/122.

Werner, F. "What Every Transportation Manager Should Know about GARVEEs." FHWA Resource Center. 2007. http://www.fhwa.dot.gov/ipd/pdfs/finance/Federal_Debt_Financing_Tools_120109.pdf.

B-9. RAILROAD REHABILITATION AND IMPROVEMENT FINANCING

The Railroad Rehabilitation and Improvement Financing (RRIF) program, directed by DOT's Federal Railroad Administration (FRA), provides loans and loan guarantees out of a $35-billion pool of revolving credit. RRIF funds can be used to "acquire, improve, or rehabilitate intermodal or rail equipment or facilities, including track, components of track, bridges, yards, buildings and shops; refinance outstanding debt incurred for the purposes listed above; and develop or establish new intermodal or railroad facilities."[54]

KEY FACTORS FOR EVALUATION

Applicability to different types of infrastructure: The use of RRIF funds is limited to the rail itself or related facilities and therefore could not fund TOD infrastructure as defined in this report. However, RRIF is an example of a financing tool that could help make TOD infrastructure projects possible by funding the transit and thereby potentially freeing up other funds that could be applied to the TOD infrastructure.

Approval requirements and legal and political considerations: RRIF loans and loan guarantees are granted under a competitive process. There are few legal or political considerations associated with this tool.

Application for strong and weak real estate markets: This tool can be applied in both strong and weak real estate markets.

Capacity and scale: RRIF loans have ranged from $2 million to $233 million.

Ease of use: Although the application process is challenging, once approved, the execution of RRIF financing is fairly straightforward. RRIF loans are typically part of a larger, complex funding and financing package that might require a dedicated administrator.

USE OF THE TOOL IN PRACTICE

The fund has made at least 33 loans to public and private entities for rail and rail facilities between 2002 and 2012.[55] Only a handful of transit-related projects have been financed by RRIF, which tends to focus on freight railroads.[56]

Example: Denver Union Station

Location: Denver, Colorado

---

[54] FRA. "Railroad Rehabilitation & Improvement Financing (RRIF) Program." http://www.fra.dot.gov/rpd/freight/1770.shtml. Accessed July 18, 2012.

[55] Ibid.

[56] Center for Transit-Oriented Development. *Financing Transit-Oriented Development: Framing the Issues and Assessing the Tools.* Prepared for Transportation for America (unpublished). June 2011.

Description: The Denver Union Station project received an RRIF loan for $152.1 million in conjunction with a $151.6 million TIFIA loan to support the development of a new multimodal station connecting light rail, commuter rail, buses, streets, and public spaces.[57] Together, the federal loans made up 64 percent of the nearly half-billion-dollar project cost. The proposed office and retail development in the station area will produce tax increment that can support local bond repayments. However, TIFIA and RRIF lowered the cost of borrowing.[58]

GETTING STARTED

The first step in the RRIF application process is a pre-application meeting with FRA staff. Then, a project sponsor completes an application and submits it to FRA. Many RRIF applicants hire an external advisor to help prepare the application. Applications are subject to an intensive vetting and due diligence process, and it can take several months before they are transmitted to the DOT Secretary's Office and the Office of Management and Budget for final decision. If accepted, a financing agreement is negotiated, and draw-downs on the loan can begin.

REFERENCES

Center for Transit-Oriented Development. *Financing Transit-Oriented Development: Framing the Issues and Assessing the Tools.* Prepared for Transportation for America (unpublished). June 2011.

Denver Union Station Project Authority. "Home page." http://www.denverunionstation.org. Accessed August 22, 2011.

FRA. "Railroad Rehabilitation & Improvement Financing (RRIF) Program." http://www.fra.dot.gov/rpd/freight/1770.shtml. Accessed July 18, 2012.

Loftus, Thomas. "The Federal RRIF Loan Program: An Option for Rail Project Financing for Public Entities." National Railroad Construction and Maintenance Association. April 2011. http://www.nrcma.org/download.cfm?ID=27751.

---

[57] Denver Union Station Project Authority. "Home page." http://www.denverunionstation.org. Accessed August 22, 2011.

[58] A more detailed description of the Denver Union Station project is in Chapter III, Section C.

C. CREDIT ASSISTANCE

C-1. CREDIT ASSISTANCE TOOLS

Federal and state agencies have developed a variety of financial tools to help local governments access credit to expedite infrastructure projects. This credit assistance can take several forms:[59]

- Bond insurance, where a local government can receive a better bond rating based on the guarantee of another agency rather than its own underlying credit.

- Credit enhancements, where federal or state funds are made available as a line of credit.

- Credit lines, a type of loan used only in the event of a shortfall in revenue for debt service.

- Loans, where a federal or state agency or program lends funds directly to a local government agency or nonprofit partner.

- Loan guarantees, where a federal or state agency agrees to cover the borrower's debt obligation if the borrower defaults.

Credit assistance improves local agencies' creditworthiness and thus lets them access better borrowing terms and lower financing costs. For example, a local agency might have access to a line of credit provided by a state or federal agency, which would help reduce the risk of default if the local agency had a temporary shortfall and otherwise would not be able to make its loan payment. This contingent loan reduces investors' risk exposure, allowing local government project sponsors to either borrow at lower interest rates or have access to the debt market that would otherwise not be possible without credit assistance. Loans from federal and state agencies can also serve as credit assistance by reducing the amount of capital borrowed from other sources, thereby reducing the risk borne by other investors and providing the capital to proceed with a project.

As described above, credit assistance can be provided by state or federal agencies and can take several forms. Examples of federal credit assistance programs relevant to providing TOD infrastructure include:[60]

- **U.S.C. Title 23, Section 129:** Under Section 129, states can use federal highway funds to make loans to local governments for eligible projects. The loans must be repaid with a dedicated, non-federal source, such as tolls, excise taxes, sales taxes, real property taxes, motor vehicle taxes, incremental property taxes, or other fees. Originally created to support toll projects, Section 129 was extended to non-toll facilities with dedicated revenue sources. The loan or credit enhancement can be for up to the maximum federal-aid share (usually 80 percent) and has fewer federal requirements than a state infrastructure bank loan. Section 129 loans have largely been superseded by TIFIA, which has dedicated funds and is direct federal aid to projects through loans, loan guarantees, or lines of

---

[59] FHWA. "Tools & Programs: Federal Credit Assistance." http://www.fhwa.dot.gov/ipd/finance/tools_programs/federal_credit_assistance/index.htm. Accessed August 22, 2011

[60] Ibid; FHWA. "Glossary." http://www.fhwa.dot.gov/ipd/glossary/index.htm. Accessed August 22, 2011.

credit. However, projects that do not meet TIFIA's $50-million threshold or other criteria can use this tool. The program is administered by state departments of transportation and FHWA.

- **Transportation Infrastructure Finance Innovation Act:** TIFIA provides federal credit assistance in the form of direct loans, loan guarantees, and standby lines of credit to finance surface transportation projects.[61]

- **State infrastructure banks:** SIBs give states the capacity to use federal highway funds to seed revolving funds that can provide loans, guarantees, lines of credit, and bond issuances for transportation projects.[62,63]

KEY FACTORS FOR EVALUATION

Applicability to different types of infrastructure: In general, credit assistance could be applicable to any type of TOD infrastructure project. In practice, most federal credit assistance tools have been designed for large surface transportation projects (e.g., toll roads and bridges). In addition, federal and state programs are typically limited to projects with a dedicated revenue source.

Approval requirements and legal and political considerations: Obtaining credit assistance from federal programs typically requires obtaining approval first from the state department of transportation and then from FHWA. Eligibility for credit assistance from an SIB varies by state depending on the bank's policies.

Application for strong and weak real estate markets: Market strength has no direct implications for credit assistance unless a pledged source for debt repayment is related to property or real estate (e.g., property taxes). If the real estate market is weak, then the borrower's default risk increases, and the terms of the loan will be less favorable.

Capacity and scale: Many of the federal credit assistance tools are designed for large projects. The Section 129 programs, however, are applicable to projects that do not meet TIFIA's larger size thresholds. Smaller projects can access loans and credit assistance tools through SIBs.

Ease of use: Credit assistance is typically part of a complex financial package that relies on several funding and financing mechanisms. Federal credit assistance often comes in the form of direct federal loans, meaning projects must comply with mandates such as federal design standards and National Environmental Policy Act requirements.

USE OF THE TOOL IN PRACTICE

Credit assistance does not appear to have been widely used for TOD infrastructure projects, but it has been used for larger transit projects that could include TOD infrastructure.

---

[61] TIFIA is described in more detail in Section C-2 of this appendix.

[62] FHWA. "Fact Sheets on Highway Provisions: State Infrastructure Bank Program." http://www.fhwa.dot.gov/safetealu/factsheets/sibs.htm. Accessed August 22, 2011.

[63] State infrastructure banks are described in more detail in Section B-7 of this appendix.

Example: Washington Dulles Corridor Metrorail Project

Location: Northern Virginia

Description: The Metropolitan Washington Airports Authority (MWAA) is constructing a 23-mile extension of the existing Metrorail system, which will be operated by WMATA (Exhibit B-4). At the end of 2011, DOT, WMATA, MWAA, and the Virginia counties of Fairfax and Loudoun entered into an agreement under which MWAA and the counties will receive credit assistance for the Dulles Corridor Metrorail Project. Through TIFIA, DOT will provide credit assistance up to $30 million for projects including parking facilities and a station.

Exhibit B-4. Construction of the Metrorail extension in Fairfax County, Virginia. Source: Stephen Barna via Flickr.com.

GETTING STARTED

The first step for most federal and state credit assistance programs is for the state department of transportation to identify a qualifying project and a local government project sponsor that could benefit from public credit assistance. The state department of transportation and the local government project sponsor determine the approximate amount of a loan or other credit assistance needed. For federal credit assistance programs, the next step is for the state to discuss the project and loan structure with the federal agency (typically FHWA). For some federal credit assistance programs, qualifying infrastructure projects are required to obtain an investment grade rating on their senior debt[64] obligations.

FHWA's Innovative Program Delivery Office can help with federal credit assistance and other programs. Although the office focuses on surface transportation projects, it could provide assistance with TOD projects. The mechanism for providing assistance varies according to the nature of the request.[65]

REFERENCES

FHWA. "Fact Sheets on Highway Provisions: State Infrastructure Bank Program." http://www.fhwa.dot.gov/safetealu/factsheets/sibs.htm. Accessed August 22, 2011.

FHWA. "Glossary." http://www.fhwa.dot.gov/ipd/glossary/index.htm. Accessed August 22, 2011.

FHWA. "Project Profiles: Washington Metro Capital Improvements Program." http://www.fhwa.dot.gov/ipd/project_profiles/dc_metro_cip.htm. Accessed August 22, 2011.

FHWA. "Tools & Programs: Federal Credit Assistance." http://www.fhwa.dot.gov/ipd/finance/tools_programs/federal_credit_assistance/index.htm. Accessed August 22, 2011.

---

[64] Senior debt is debt that takes priority over other debt securities. If the debtor goes bankrupt, senior debt must be repaid before other creditors receive payment.

[65] For more information see: FHWA. "How the IPD Office Does Business...Technical Assistance." http://www.fhwa.dot.gov/ipd/how_business/technical_assistance.htm. Accessed July 2011.

C-2.   TRANSPORTATION INFRASTRUCTURE FINANCE AND INNOVATION ACT

The Transportation Infrastructure Finance and Innovation Act (TIFIA) was created by Congress in 1998 to provide federal credit assistance (e.g., secured loans, loan guarantees, or lines of credit) for projects larger than $50 million that face financing challenges due to their size or complexity.[66] TIFIA assistance can be applied toward a project's capital costs and operations and maintenance.[67]

TIFIA is authorized for $750 million for 2013 and $1 billion 2014, up from $122 million in 2011.[68] Each TIFIA dollar can provide credit assistance for up to $10 and leverage $30 in other investment.[69] The 2012 transportation reauthorization bill increased the size of available TIFIA loans from a maximum of 33 percent to a maximum of 49 percent of eligible project costs, which include expenses for project development, right-of-way acquisition, procurement, construction, and capitalized interest on the senior debt. The bill also requires TIFIA to adopt a rolling basis for applications and make awards based on availability of funds, which means that there would be no deadlines for submissions.[70]

KEY FACTORS FOR EVALUATION

Applicability to different types of infrastructure: TIFIA supports surface transportation projects larger than $50 million, including transit projects, passenger and freight rail, and intercity passenger facilities. Eligible projects include design and construction of stations, tracks, and other infrastructure and purchase of transit and intercity passenger rail vehicles.[71]

For real property, such as transit stations, the costs associated with property must be physically and functionally related to the transportation project to be considered eligible costs; a parking facility could be eligible if its purpose is to support transit use. Real estate development costs, including the acquisition cost of land outside the immediate right-of-way, would not be eligible.

Approval requirements and legal and political considerations: All assistance is awarded based on a project's creditworthiness and the availability of funds. In addition projects must have a dedicated non-federal revenue source, be included in the state Transportation Improvement Program, and receive an investment-grade rating on the senior debt.

Capacity and scale: To be considered for TIFIA funding, a project must exceed $50 million. However, the project can include several components, such as a station, a parking lot (if the parking is supporting transit use), and tracks, to reach or exceed the $50-million threshold.

---

[66] de la Pena, P.; Caplicki, E. V.; and Santiago, S. J. "2010 Transportation Infrastructure Year in Review." Nossaman LLP. February 17, 2011.

[67] FHWA. "TIFIA Defined." http://www.fhwa.dot.gov/ipd/tifia/defined/. Accessed August 23, 2012.

[68] FHWA. "Transcript and chatroom comments from MAP-21 Webinar." http://www.fhwa.dot.gov/ipd/tifia/public_outreach/map21_outreach1_0812.htm. Accessed August 23, 2012.

[69] FHWA. "TIFIA." http://www.fhwa.dot.gov/ipd/tifia/. Accessed July 18, 2012.

[70] Kessler, F. W. and Denton, P. W. "MAP-21: Surface Transportation Reauthorization Ushers in Significant Changes to TIFIA." Nossaman LLP. July 6, 2012.

[71] FHWA. *TIFIA Program Guide.* 2012.

Timing and lifecycle Projects receiving TIFIA assistance have a maximum repayment period of 35 years after a project's substantial completion. The TIFIA interest rate is a fixed rate based on the U.S. Treasury rate.

At DOT's discretion, debt service can be deferred for up to five years after substantial completion. DOT can also structure a debt service schedule that aligns repayment with projected cash flows. This schedule might include deferring partial interest and principal repayments beyond the five-year, post-construction period as needed. Projects are not entitled to debt service deferral; DOT can evaluate each project's economics to determine an appropriate repayment schedule.

Ease of use: TIFIA assistance under the most recent authorization will be awarded on a first-come, first-served basis so applicants able to submit requests early in the process will be most competitive.

USE OF THE TOOL IN PRACTICE

TIFIA assistance has supported TOD components such as parking facilities, transit stations, and rail. Examples include the San Francisco Transbay Transit Center, Miami Intermodal Terminal, and Puerto Rico Commuter Train (Tren Urbano). The San Francisco Transbay Transit Center was the first TIFIA loan secured by value capture[72] revenue from property taxes on surrounding TOD. It presents an innovative approach in TOD financing combining both federal sources and value capture.

Example: San Francisco Transbay Transit Center

Location: San Francisco, California

Description: The Transbay Transit Center Project will replace the existing Transbay Terminal with a new multimodal transportation center that can accommodate nine transportation systems. Construction began in 2008 and is scheduled to be completed in 2017.[73] The project has three parts: replacing the outdated terminal; extending the Caltrain rail line 1.3 miles into the new terminal; and redeveloping the area around the Transbay Transit Center with new homes (35 percent of which will be affordable), parks, and shops. Phase 1 of the project includes the transit center building and part of the Caltrain rail extension. Phase 2 will complete the rail extension; the funding and financing sources for this phase have not yet been secured.[74]

The funding and financing for Phase 1 come from multiple sources, including sales taxes (8 percent), toll bridge revenue (30 percent), state funds (2 percent), land sales (36 percent), SAFETEA-LU and FTA Section 1601 (5 percent), TIFIA loan (14 percent), and other sources (5 percent). The TIFIA loan is secured by a senior lien on project revenue, which include dedicated tax increment revenue from land

---

[72] For more information on value capture mechanisms, see Section I.E in this appendix.

[73] Transbay Transit Center. "Project Schedule." http://transbaycenter.org/construction-updates/project-schedule. Accessed July 18, 2012.

[74] FHWA. "Project Profiles: Transbay Transit Center." http://www.fhwa.dot.gov/ipd/project_profiles/ca_transbay_transit.htm. Accessed July 25, 2011.

sold and developed in the state-owned parcels surrounding the transit center, and a commitment of passenger facilities charges from the transit center's initial primary tenant, AC Transit. [75]

GETTING STARTED

Applicants must submit a letter of interest, and if invited to submit a formal TIFIA application, they must pay a non-refundable application fee. For projects that enter credit negotiations, sponsors must pay a transaction fee to DOT regardless of whether the loan closes.

REFERENCES

de la Pena, Patricia; Caplicki, Edmund V.; and Santiago, Simon J.. "2010 Transportation Infrastructure Year in Review." Nossaman LLP. February 17, 2011. http://www.nossaman.com/7749.

FHWA. "Project Profiles: Transbay Transit Center." http://www.fhwa.dot.gov/ipd/project_profiles/ca_transbay_transit.htm. Accessed July 25, 2011.

FHWA. "TIFIA." http://www.fhwa.dot.gov/ipd/tifia/. Accessed July 18, 2012.

FHWA. "TIFIA Defined." http://www.fhwa.dot.gov/ipd/tifia/defined/. Accessed August 23, 2012.

FHWA. *TIFIA Program Guide.* 2012. http://www.fhwa.dot.gov/ipd/pdfs/tifia/tifia_program_guide_0612.pdf.

FHWA. "Transcript and chatroom comments from MAP-21 Webinar." http://www.fhwa.dot.gov/ipd/tifia/public_outreach/map21_outreach1_0812.htm. Accessed August 23, 2012.

Kessler, Fredric W. and Denton, Peter W. "MAP-21: Surface Transportation Reauthorization Ushers in Significant Changes to TIFIA." Nossaman LLP. July 6, 2012. http://www.nossaman.com/SignificantChangestoTIFIA.

Neaher, Ned. "TIFIA and private activity bonds: Additional financing sources for US PPP transportation projects." Transportation Finance Review. October 2007.

Podkul, Cezary. "TIFIA oversubscribed again." PEI Media Limited. March 25, 2011. http://www.infrastructureinvestor.com/Article.aspx?article=60219.

---

[75] Ibid.

D. EQUITY SOURCES

D-1. PUBLIC-PRIVATE PARTNERSHIPS

A public-private partnership (P3) is defined as "a contractual agreement between a public agency (federal, state, or local) and a private-sector entity. Through this agreement, the skills and assets of each sector (public and private) are shared in delivering a service or facility for the use of the general public. In addition to the sharing of resources, each party shares in the risks and rewards in the delivery of the service and/or facility."[76] An infrastructure P3 has the following key elements:

- A long-term contract between a public-sector party and a private-sector party.

- Design, construction, financing, and operation of public infrastructure by the private-sector party.

- Payments over the life of the P3 contract to the private sector for the use of the facility, made either by the public sector or by the general public as users of the facility.

- The facility remaining in public-sector ownership or reverting to public-sector ownership at the end of the P3 contract. [77]

In a typical P3, the private entity provides the capital cost to finance the project. If the project generates enough revenue to cover its construction and operation costs, the P3 will commonly use a concession lease where the private partner makes an upfront or ongoing payment to the public partner in exchange for developing (if required), financing, operating, and maintaining the asset. Under this approach, the private partner would collect the revenue generated by the asset. Examples of this type of P3 arrangement are parking facilities, toll roads, airports, and ports.

If the project does not generate any revenue (e.g., library, school, parks, etc.) or does not generate enough revenue to cover its capital and operation costs, then the P3 can use an availability payment approach. Under this approach, the private partner designs, builds, finances, and maintains a publicly owned facility. The private partner is paid back by the public partner during operations through periodic payments based on meeting standards for the physical conditions of the facility established by the public partner. If the facility does not comply with the standards, then the public partner can apply penalties or deductions to the periodic payment.

Procuring infrastructure projects using a P3 approach has several potential benefits.

- P3 capital costs are spread over the life of the asset, allowing government to proceed with projects that it might not otherwise be able to afford with currently available funds.

- The government transfers certain risks to the private-sector entity. P3s allocate a project's risks to the party best suited to manage them. Typically, design, construction, and operation risks are the

---

[76] National Council for Public-Private Partnerships. "How PPPs Work." http://www.ncppp.org/howpart/index.shtml. Accessed July 18, 2012.

[77] Yescombe, E.R. *Public-Private Partnerships—Principles of Policy and Finance*. Elsevier Ltd. 2007.

responsibility of the private partner, while right-of-way acquisition and force majeure events are the responsibility of the public partner.

- Involving the private sector in procurement and development of P3 projects increases the likelihood that only economically viable projects will proceed. Furthermore, P3 projects can take advantage of the efficiency and innovation offered by the private sector.

KEY FACTORS FOR EVALUATION

Applicability to different types of infrastructure: A P3 can be used for all types of public infrastructure. Projects could be new construction or upgrades to existing assets.

Approval requirements and legal and political considerations: Implementation of P3 projects varies greatly across states. In general, each state enacts its own legislation to authorize, prohibit, or limit P3 transactions. Authority for a P3 transaction can also be drawn from general statutory powers granted to state or local government entities.

Exhibit B-5 summarizes the P3 state legislation for Colorado, Georgia, Illinois, and Utah.[78]

State	P3 Enabling Legislature?	Limited to
Colorado	Yes	Correctional facilities, college saving accounts, administration of water bank, transportation projects (rail, highways), Major League Baseball stadium
Georgia	Yes	Transportation projects that generate "greatest gains in congestion mitigation or promotion of economic development" and water resources projects
Illinois	Yes	Riverdale brownfield redevelopment, bridges, parking garage, lottery system, highway, toll highway, tunnel, intermodal facility, intercity or high-speed passenger rail, or other transportation facility or infrastructure.
Utah	Yes	Tollway facilities

Exhibit B-5. P3 Enabling legislation.

In addition to satisfying the applicable P3 enabling legislation, a P3 candidate project might also need to undergo several rounds of hearings and approvals from various agencies and review boards, depending on state and local rules.

Application for strong and weak real estate markets: P3 can be affected by the real estate market conditions if the revenue stream (or a significant share of it) to pay for the facility to be built and maintained by the private sector comes from property leases or property taxes. For example, if the availability payment will come from property taxes, private investors will perceive higher risk in a weak real estate market and will require additional guarantees (e.g., a longer contract term) or a higher rate of return to recover their investment.

---

[78] Pikiel, M. E. and Plata, L. "A Survey of PPP Legislation Across the United States." *Global Infrastructure.* 2008. 1:52-65.

Capacity and scale: P3s are typically large, complex projects such as transportation or social infrastructure (e.g., schools, hospitals, or libraries). Smaller projects might need to be bundled or included as part of a larger P3 project to attract private investment. These bundled projects could involve parks; streetscaping; road, bicycle, or pedestrian improvements; sewer, water, storm drain, and other utilities; or parking.

Timing and lifecycle: Regardless of the specific approach (availability payment versus concession lease), the private partner, or "concessionaire," is responsible for raising its own financing. A concessionaire can generally generate the maximum amount of financing by using both debt and equity, which typically consists of 10 to 20 percent equity, over a long-term concession (typically 30 years or more). For funds that are to be used on qualifying capital expenditures, a concessionaire would be able to access many of the same debt markets as the public sector, including tax-exempt bonds (in the form of qualifying private activity bonds that are available for such projects) and federal programs.

Ease of use: While P3s theoretically can be used for projects of all sizes and types, procuring a P3 project is a complex process that involves multiple advisors to coordinate legal, technical, and financial issues, which can result in a longer, more expensive procurement process. P3 procurement costs can reach 5 to 10 percent of a large project's capital cost, and the procurement costs for smaller projects often are greater.

USE OF THE TOOL IN PRACTICE

Most of the transportation projects built through P3s are highway projects. Though the example here is not a TOD project, it does offer insights into an infrastructure project delivered as a P3 using availability payments. This P3 structure can be used to build facilities such as transit stations or projects supporting TOD such as street enhancements, bike paths, or parks.

Example: Long Beach Courthouse

Location: Long Beach, California

Description: Long Beach Courthouse was the first P3 deal in the state of California and the first in the United States to use an availability payment approach for a social infrastructure project. The project consists of the construction of a new building of approximately 500,000 square feet that will house 31 courtrooms, offices of county justice agencies, and commercial space compatible with courthouse uses. The project also includes renovation of the nearby parking structure. The Long Beach Judicial Partners consortium will design, build, and then maintain the facility for 35 years. In December 2010, the consortium secured a project financing package consisting of long-term equity (10 percent) and debt (90 percent), with a total investment of just under $500 million.[79]

In exchange for these services, the consortium will be paid through an annual service fee or availability payment made by the state's Administrative Office of the Courts, which owns the land and the building. The payments will start once the construction is completed and continue for a 35-year operation period at which time the Administrative Office of the Courts will take over control of the building. The service

---

[79] Barandiaran, I. "Social Infrastructure Public-Private Partnerships—Helping Government Overcome Budget Constraints to Deliver Public Services." *Perspectives on Real Estate Newsletter.* Summer 2011.

fee payments are linked through potential deductions to specific availability and performance indicators, including the availability of fully functioning court rooms for their intended use every day of the year. The consortium thus has an incentive to complete the construction on time and on budget, so that service fee payments start as scheduled, and to operate and maintain the building in good condition to minimize or avoid any potential payment deductions.[80]

GETTING STARTED

Before a project sponsor decides to use a P3 to deliver a project, it will need a thorough analysis of the project's legal, technical, and financial feasibility. This analysis requires:

- Identifying which project or projects are suitable for a P3 under existing P3 enabling legislation.

- Completing environmental clearance, if needed.

- Performing a "value for money" analysis comparing the benefits and costs of P3 and traditional public procurement methods. The value-for-money analysis will determine whether a P3 would save the project sponsor enough money and provide potential investors with a high enough rate of return.

Having experienced advisors is often critical for a project's success. Each P3 is structured differently depending on its needs and market conditions. Typically, the project sponsor has technical, legal, and financial advisors that guide it through the transaction from project screening to preparing bid documents, evaluating proposals, and negotiating with the preferred bidder.

REFERENCES

Barandiaran, Ignacio. "Social Infrastructure Public—Private Partnerships-Helping Government Overcome Budget Constraints to Deliver Public Services." *Perspectives on Real Estate Newsletter.* Summer 2011. http://www.pillsburylaw.com/index.cfm?itemID=40191&pageID=34.

National Council for Public-Private Partnerships. "How PPPs Work." http://www.ncppp.org/howpart/index.shtml. Accessed July 18, 2012.

Pikiel, Michael E. and Plata, Lillian. "A Survey of PPP Legislation Across the United States." *Global Infrastructure.* 2008. 1:52-65.

Yescombe, E.R. *Public-Private Partnerships—Principles of Policy and Finance.* Elsevier Ltd. 2007.

---

[80] Ibid.

D-2.   INFRASTRUCTURE INVESTMENT FUNDS

As the need for investment in infrastructure continues to grow, private financing for infrastructure projects has developed around the world though infrastructure investment funds.[81] While already established in Australia, Canada, and Europe, treating infrastructure as an asset class is still relatively new in the United States. Infrastructure investment funds have supported projects in a broad range of sectors such as transportation (e.g., toll roads, airports, ports, and transit), regulated utilities (e.g., water and power), cable and wireless communication, and social infrastructure (e.g., schools, hospitals, public and military housing, and civic buildings).[82]

An investment fund is a pool of funds collected from many investors to invest in infrastructure, often in the form of a public-private partnership. An infrastructure investment fund can be the financing tool that pays for a public project's capital cost under a public-private partnership. Some pension funds have increased their allocation to alternative investments like infrastructure in an attempt to both reduce risk through diversification and generate higher risk-adjusted returns.[83]

Similar to other types of public-private partnerships, infrastructure investment funds seek projects with stable, predictable, and long-term income streams. While infrastructure investment funds have not been widely used in the United States, there are examples of private investment in infrastructure through public-private partnerships in states or localities where enabling legislation exists. This tool has not been applied to TOD-related infrastructure as defined in this report, although public-private partnerships have been used to finance transit. However, infrastructure investment funds could invest in revenue-generating, TOD-related infrastructure projects like parking, utilities, and toll roads if the project generated sufficient returns to be an attractive investment.

In 2012 the Chicago City Council passed an ordinance to create the Chicago Infrastructure Trust, a nonprofit entity that the city will use as a financing tool for a planned $1 billion in infrastructure projects. Under the Chicago Infrastructure Trust, the city is making agreements with private investment and financing firms to provide financing for infrastructure projects that have a defined revenue stream or the potential for a fee or surcharge that would pay back the investment.

KEY FACTORS FOR EVALUATION

Applicability to different types of infrastructure: Infrastructure investment funds in the United States have primarily invested in revenue-generating infrastructure, most typically toll roads or bridges and utilities tied to a steady revenue stream. In Europe and Canada, infrastructure funds have solid experience investing in non-revenue-generating projects such as civic and social infrastructure.

Approval requirements and legal and political considerations: Private investment in public infrastructure typically requires state-enabling legislation for public-private partnerships. In some cases, political opposition to private involvement in providing public infrastructure has arisen because of assumptions that the private sector would be motivated by profit and not necessarily the public good.

---

[81] Inderst, G. *Pension Fund Investment in Infrastructure.* OECD Publishing. 2009. p. 34.

[82] Deloitte. *REITs and infrastructure projects. The next investment frontier?* 2010. p. 1.

[83] Inderst, G. *Pension Fund Investment in Infrastructure.* OECD Publishing. 2009. p. 3.

Ease of use: Attracting investment from investment funds to infrastructure projects requires a certain level of project readiness such as environmental clearance and secure cash flows (e.g., tolls, lease payments, or public guarantees), often with inflation-protected returns. In addition, the transaction costs of public-private partnerships can be high because this type of financing is typically very complicated. A public agency could have difficulty determining if it is getting a good deal and must rely on a group of legal, financial, insurance, and technical advisors.

Capacity and scale: Infrastructure investment funds generally focus on medium to large projects. Relatively small projects might need to be packaged to attract private investment.

USE OF THE TOOL IN PRACTICE

The Chicago Skyway toll road illustrates how a partnership involving an infrastructure investment fund might work. This tool could be applied in a TOD context if the project sponsor could identify an infrastructure project (e.g., parking, utilities, or toll roads) that generated sufficient revenue to be attractive to investors.

Example: Chicago Skyway Toll Road

Location: Chicago, Illinois

Description: In the Chicago Skyway transaction, completed in 2005, the city of Chicago granted a 99-year lease to two private infrastructure investment groups to operate, maintain, manage, rehabilitate, and toll the road. The transaction raised $1.8 billion in revenue for Chicago. A consortium formed under the deal is responsible for all operations and maintenance costs of the skyway and has the right to all toll and concession revenue. This agreement was the first long-term lease of an existing toll road in the United States.[84]

For another example of an investment fund provided financing related to transit, see the Eagle public-private partnership example in Section D-1 of this appendix.

GETTING STARTED

To attract private investors, a project sponsor needs a thorough analysis of an infrastructure project's legal, technical, and financial feasibility. Likely steps in the analysis include:

- Identifying which project or projects are suitable for a public-private partnership under existing legislation.

- Obtaining environmental clearance.

- Performing a value-for-money analysis to determine whether a public-private partnership would provide enough cost savings to the project sponsor and a high enough rate of return to potential investors.

---

[84] FHWA. "Project Profiles: Chicago Skyway." http://www.fhwa.dot.gov/ipd/project_profiles/il_chicago_skyway.htm. Accessed August 22, 2011.

REFERENCES

Deloitte. *REITs and infrastructure projects. The next investment frontier?* 2010. http://www.deloitte.com/assets/Dcom-UnitedStates/Local%20Assets/Documents/MA/us_ma_Infrastructure%20REITS_040210.pdf.

FHWA. "Project Profiles: Chicago Skyway." http://www.fhwa.dot.gov/ipd/project_profiles/il_chicago_skyway.htm. Accessed August 22, 2011.

Inderst, Georg. *Pension Fund Investment in Infrastructure.* OECD Publishing. 2009. http://www.oecd.org/insurance/privatepensions/42052208.pdf.

E. VALUE CAPTURE MECHANISMS

E-1. DEVELOPER FEES AND EXACTIONS[85]

Developer fees and exactions include:

- Impact fees, which include system development charges and connection or facility fees, and

- Negotiated exactions and agreements.

IMPACT FEES

Development impact fees, system development charges, and connection or facility fees are charges assessed on new development to defray the cost to the jurisdiction of extending public services to the development and cannot be used to fund existing deficiencies. The fees are generally collected once and are used to offset the cost of providing public infrastructure such as streets and utilities. Many jurisdictions have transportation impact fees that include an allocation for transportation improvements, but most are focused on roads.[86] Broward County, Florida, and San Francisco use impact fees to pay for transit service.

Although fee eligibility and structure vary by state, in general impact fees must be adopted based on findings of reasonable relationships between the development paying the fee, the need for the fee, and the use of fee revenue. Local governments can allow credits and reimbursements[87] for capital projects funded by an impact fee that are constructed privately by developers and dedicated to a public entity. Depending on the fee program's guidelines, a development project could choose to dedicate land or make certain improvements and receive a credit against the impact fee due.

NEGOTIATED EXACTIONS AND AGREEMENTS

Direct contributions from developers can also help pay for infrastructure to accommodate new development. Jurisdictions and developers often negotiate to obtain desired improvements in exchange for development rights. The extent to which a new project can contribute to the provision of infrastructure depends on many factors, including the anticipated revenue from development, construction costs, lot size and configuration, and parking ratios. All of these factors vary depending on the form and timing of development, and therefore the amount of public benefits that can be provided through developer agreements is unpredictable and has to be negotiated.

---

[85] This tool description is based on previous work conducted by the Center for Transit-Oriented Development and Strategic Economics, including the report, *Capturing the Value of Transit*.

[86] Center for Transit-Oriented Development. *Capturing the Value of Transit*. Prepared for FTA. 2008. p. 30.

[87] A "credit" is the amount counted against the developer's fee obligation. A "reimbursement" is the amount that exceeds the developer's fee obligation.

KEY FACTORS FOR EVALUATION

Applicability to different types of infrastructure: Impact fees can be used to fund any type of infrastructure for which a local government can demonstrate an immediate increase in need due to the new development. Depending on the context, any of the TOD infrastructure discussed in this report could be included.

Approval requirements and legal and political considerations: Fees must be based on a connection between the impacts of new development and the amount of the fee. Local governments would likely need to take legislative action to establish impact fees, and sometimes state legislative action is also needed.

Application for strong and weak real estate markets: This tool relies on new development occurring to realize any revenue and thus is most applicable to strong real estate markets.

Capacity and scale: The amount of money that an impact fee can generate is directly tied to the estimated impacts of the new development. If the need for a new facility, whether it is a bicycle rack, a road, or a park, is triggered by development, the development could, in theory, be responsible for the entire cost of the facility.

Ease of use: Impact fees require administration, monitoring the collection and designation of the fee revenue, and documenting the use for approved projects.

OTHER LIMITATIONS OF THE TOOL

Because development fees can only be used on a pay-as-you-go basis, they are difficult, if not impossible, to bond against.

USE OF THE TOOL IN PRACTICE

Example: Boston Linkage Fees for Affordable Housing and Land Acquisition

Location: Boston, Massachusetts

Description: The city of Boston instituted a development fee for any large housing and commercial development that requires a zoning change. The city created a Neighborhood Housing Trust to manage the housing linkage funds and a Neighborhood Jobs Trust to manage the jobs linkage funds. After the fee was established, it faced a legal challenge, so the city submitted a home rule petition to the Massachusetts legislature that resulted in legislative authorization for Boston's linkage program. The Boston Zoning Commission later incorporated the fee into Boston's zoning code.[88]

Developers required to pay the fee enter into an agreement with the Boston Redevelopment Authority to confirm the payment of linkage fees. The Neighborhood Housing and Neighborhood Jobs Trusts, in conjunction with the redevelopment authority, use the revenue to fund affordable housing and jobs

---

[88] City of Boston. *Neighborhood Housing Trust.* 2011.

programs. For example, the redevelopment authority has used the funding to provide no-interest loans for land acquisition to local community development corporations to develop affordable housing.

Example: New Quincy Center Infrastructure Reimbursement Agreement

Location: Downtown Quincy, Massachusetts

Description: The city of Quincy has entered into a long-term partnership with StreetWorks LLC to replace all the existing infrastructure in the city's downtown as part of StreetWorks' plan to redevelop 50 acres in the city center with a mix of retail and entertainment uses, health facilities, educational institutions, and housing.[89] The project is next to a Red Line subway stop, Quincy Center Station. StreetWorks will raise private funding, backed by city bond guarantees, to build new utilities, roads, sidewalks, parking garages, and open space. Once the new infrastructure is complete and producing revenue, the city will reimburse the developer by purchasing the infrastructure from StreetWorks by selling general obligation bonds.[90]

GETTING STARTED

Although the process of establishing an impact fee varies by state, local governments typically begin the process by commissioning a "nexus study" to establish a direct connection between the use and size of the fee and the impact of new development.

REFERENCES

Center for Transit-Oriented Development. *Capturing the Value of Transit*. Prepared for FTA. 2008. http://reconnectingamerica.org/resource-center/books-and-reports/2008/capturing-the-value-of-transit-3.

City of Boston. *Neighborhood Housing Trust*. 2011. http://www.cityofboston.gov/dnd/pdfs/NHT.pdf.

City of Quincy, Planning and Community Development Department. "Downtown Revitalization." http://www.quincyma.gov/Government/PLANNING/DowntownRevitalization.cfm. Accessed August 22, 2011.

Diesenhouse, Susan. "Rebuilding Downtown From the Ground Up." *New York Times*. April 5, 2011. http://www.nytimes.com/2011/04/06/realestate/commercial/06quincy.html.

Duncan Associates. "Welcome to ImpactFees.com." 2008. http://www.impactfees.com. Accessed August 22, 2011.

Langdon, Philip. "Massachusetts City Aims for a Downtown Remake." *New Urban Network*. August 1, 2011. http://newurbannetwork.com/article/massachusetts-city-aims-downtown-remake-15039.

---

[89] City of Quincy, Planning and Community Development Department. "Downtown Revitalization." http://www.quincyma.gov/Government/PLANNING/DowntownRevitalization.cfm. Accessed August 22, 2011.

[90] More information about the New Quincy Center is in Chapter III, Section E.

E-2. SPECIAL DISTRICTS[91]

Special districts, which can include benefit assessment districts, business improvement districts, business improvement areas, business revitalization zones, community improvement districts, local improvement districts, special services areas, and special improvement districts, are formed to include a geographical area in which property owners or businesses agree to pay an assessment to fund a proposed improvement or service from which they expect to directly benefit.[92] Special districts are commonly used to fund infrastructure such as sewer, water, utilities, or streets but can also be used to fund services such as police, fire protection, or transit.[93] Special districts can be used either for pay-as-you-go improvements or to finance the issuance of bonds backed by the assessment revenue. When used as a financing tool, special districts tend to be less risky to the local government than many other financing tools because the risk is transferred to individual property owners.

Laws governing the use of special districts vary by state, but in most cases a special district requires a majority vote from property owners to be enacted, and in some cases a two-thirds vote is required. The amount of the assessment must be directly related to the cost of the improvement and the expected benefit to the property owner. Once passed, property owners in the district pay an additional tax or fee to pay for the service or improvement in the desired timeframe or to finance a debt obligation in accordance to the property's proportional share of the benefit. The individual property owner's tax or fee can be lower if the district encompasses a large area or is financed over a long time period. However, special districts can be more difficult to implement across larger areas, especially across multiple jurisdictions.[94]

Special districts are considered a value capture tool because they capture the value (or benefit) generated by an improvement or service to provide funding for the improvement or service. For some types of improvements, such as streetscape enhancements, commercial properties are assumed to benefit more directly, and residential properties are exempted from the assessment.

KEY FACTORS FOR EVALUATION

Applicability to different types of infrastructure: Assessments from special districts can be used to fund infrastructure that does not generate revenue, so the tool is applicable to a wide variety of circumstances. However, there must be a clear benefit to property owners who will be paying the assessment. Special assessments can finance construction of stations, and in some places, assessments have been used to help fund the transit itself in addition to the infrastructure around a station.

---

[91] This tool description is based on previous work conducted by the Center for Transit-Oriented Development and Strategic Economics, including the report, *Capturing the Value of Transit*.

[92] Another common type of service district is a parking district, an entity that manages parking in a downtown or other geographic area. Parking district revenue typically comes from charges for parking or in lieu fees rather than from special assessments. However, like other special districts, parking districts can finance improvements or services (in this case, new parking facilities or operations and maintenance of existing parking facilities) by bonding against projected future revenue.

[93] Center for Transit-Oriented Development. *Capturing the Value of Transit*. Prepared for FTA. 2008. p. 21.

[94] Ibid. pp. 21-23.

Approval requirements and legal and political considerations: Laws governing their use vary by state, but special districts typically have prerequisites for use, including local legislation to create a new district and voter approval. Assessments can be politically infeasible if property owners already feel burdened by other property taxes and assessments.

Application for strong and weak real estate markets: This financing tool does not totally rely on new development and is therefore applicable in strong and weak real estate markets. The ability to gain approval for a new assessment and the amount of revenue generated, however, often depend on the potential for new development. Although special districts are designed to capture the value conferred to existing properties by an improvement, they are easiest to implement in an area where a few property owners will be able to take advantage of a significant development opportunity. Property owners who plan to sell, develop, or redevelop their land are likely to receive more immediate benefits than other property owners from the enhanced value conferred on their properties, so they might be more willing to participate in a special district. In some cases, a special district could make development possible where it otherwise would not be.[95]

Capacity and scale: The amount of money that an assessment can raise depends on the rate that property owners are able and willing to pay, the number of property owners who are willing to participate, and the amount of new development that occurs.

Ease of use: While special districts are appealing as a value capture tool because they can align taxes closely with expected benefits, it can nevertheless be challenging to convince property owners to pay higher taxes.[96] Once established, assessments or taxes from special districts are typically collected with property taxes. Special districts might require regular renewal, and such uncertainty can cause difficulties in bonding against assessments.

OTHER LIMITATIONS OF THE TOOL

In some places only commercial properties can be assessed.

USE OF THE TOOL IN PRACTICE

Special districts have been widely used for TOD infrastructure. Typical items financed include street paving; curbs; sidewalks; street lighting; utilities, including water lines, storm and sanitary sewers, and plant expansions; parks and open space; and off-street parking. Because assessments do not need to be tied to revenue-generating infrastructure, they are particularly useful for streetscaping and other beautification projects that provide benefits to an entire district.

Example: White Flint Special Taxing District

Location: Montgomery County, Maryland (White Flint Metro Station)

---

[95] Ibid. p. 23.

[96] Ibid. p. 23.

Description: In October 2010, the Montgomery County Council approved a Special Taxing District for the White Flint Sector Plan area that is authorized to levy an ad valorem property tax.[97] The special assessment will fund the reconstruction of a major arterial as a walkable boulevard, the construction of a grid of public streets, and other infrastructure to support the redevelopment of the White Flint area as a mixed-use, transit-oriented district. The anticipated funding sources include a special assessment from the White Flint Special Taxing District (projected to pay for 63 percent of the infrastructure costs), TIF (30 percent), and an impact tax (7 percent).[98]

GETTING STARTED

The process for creating a special district varies from one jurisdiction to another, but in general the process starts with property owners or businesses petitioning the local government to create the special district. Then, the local government determines whether a majority of property owners or businesses approve the creation of the special district. Finally, the local government enacts legislation creating the special district and assessment. In some places, state legislation will be required to grant local governments the authority to create special districts.

REFERENCES

Center for Transit-Oriented Development. *Capturing the Value of Transit.* Prepared for FTA. 2008. http://reconnectingamerica.org/resource-center/books-and-reports/2008/capturing-the-value-of-transit-3.

Montgomery County Planning Department. "White Flint: North Bethesda's Urban Center." 2011. http://www.montgomeryplanning.org/community/whiteflint.

---

[97] Montgomery County Planning Department. "White Flint: North Bethesda's Urban Center." 2011. http://www.montgomeryplanning.org/community/whiteflint.

[98] More information about the White Flint Sector Plan is in Chapter III, Section F.

E-3.   TAX INCREMENT FINANCING[99]

Tax increment financing (TIF) allows the public sector to "capture" growth in property taxes (or sometimes sales taxes) from new development and increasing property values. Depending on the state, TIF can be used for individual projects or within a district. TIF works differently according to the laws in each state (except Arizona, where it is not permitted), but typically it is geared to capture the increase in property values that occurs in a designated area over a base. Tax increment is collected for a set period, usually between 15 and 30 years. It can be used either on a pay-as-you-go basis over time or can be bonded against to provide an upfront source of revenue. The most common uses of TIF are for environmental cleanup, land assembly, or local infrastructure.[100]

TIF districts must meet special criteria (usually blight conditions) to qualify. TIF financing has commonly been used to help pay for major development initiatives or infrastructure investments that catalyze private investment and increase property values. The 2008 downturn in the real estate market and constriction of real estate investment capital have made new TIF districts less viable.[101]

While the purpose of TIF is usually to encourage new development and to help revitalize distressed neighborhoods, some states are considering allowing TIF to be used for transit funding. Pennsylvania passed the Transit Revitalization Investment District (TRID) legislation in 2004 to foster integrated planning and implementation strategies for transit station areas. In a TRID, TIF can be used to fund both transit and other station-area needs. TRID has not yet been used as a financing tool in Pennsylvania, in large part due to the weak economy.[102]

TIF districts are one of the most powerful value capture tools because they capitalize on increases in property values, including the value of new development, in an entire district. As illustrated in the example in this section, Dallas created a TIF district across multiple station areas, which allows revenue generated in one station area to be deployed in another.

KEY FACTORS FOR EVALUATION

Applicability to different types of infrastructure: TIF revenue can be applied to infrastructure that does not generate revenue. Typical items financed include street improvements; sidewalks; street lighting; utilities, including water lines, storm and sanitary sewers, and plant expansions; parks and open space; and off-street parking.

Approval requirements and legal and political considerations: Laws regulating the use of TIF vary from state to state, but in most places establishing a TIF district requires certification that the area is blighted and redevelopment is necessary. Local legislative action is necessary to establish the district.

---

[99] This tool description is based on previous work conducted by the Center for Transit-Oriented Development and Strategic Economics, including the report, *Capturing the Value of Transit.*

[100] Center for Transit-Oriented Development. *Capturing the Value of Transit.* Prepared for FTA. 2008. p. 24.

[101] Center for Transit-Oriented Development. *Transit-Oriented Development Strategic Plan/Metro TOD Program: Detailed Recommendations and Background.*

[102] Center for Transit-Oriented Development. *Transit Revitalization Investment Districts (TRID): Opportunities and Challenges for Implementation.* 2011.

Establishing new TIF districts involves significant political considerations because future tax revenue is diverted from existing uses to redevelopment.

Application for strong and weak real estate markets: TIF is designed to capture value from new development and requires development activity to be effective.

Capacity and scale: The amount of tax increment that a TIF district generates, and thus the size of projects that the district can fund, depends on the size of the district and the share of tax increment that it captures.

Ease of use: Implementing and administering a TIF district is a complex task. Typically a local government would form a commission to administer the district.

OTHER LIMITATIONS OF THE TOOL

Some states restrict the total amount of assessed value that can be in a TIF district. Also, in some states, it is more difficult or not permitted to bond on the revenue because it is not backed by the full faith and credit of the government.

USE OF THE TOOL IN PRACTICE

TIF has been used extensively to pay for public improvements necessary to support TOD.

Example: City of Dallas Multistation Tax Increment Financing

Location: Eight Dallas Area Rapid Transit (DART) stations

Description: In 2008, the city of Dallas approved a series of new TIF districts that surround eight DART stations and total 559 acres.[103] The districts are the result of collaboration between the city, DART, Southern Methodist University, Prescott Realty Group, and other local partners. TIF revenue will be used for public infrastructure to support new development and enhance connectivity in station areas. TIF was considered an especially appealing alternative for DART and the city because real estate market conditions and community needs vary greatly among the different station areas.[104]

GETTING STARTED

Local governments typically establish TIF districts, but the process varies according to each state's enabling legislation. Depending on the state, the first step might be a blight analysis to determine that redevelopment of the area is required. Then, the local government might need to designate an area as either a redevelopment or economic development district and prepare a redevelopment plan that describes the activities needed, projects to be pursued, and expected uses of the TIF revenue.

---

[103] City of Dallas Economic Development. "Transit Oriented Development (TOD) Tax increment Financing (TIF) District." http://www.dallas-ecodev.org/incentives/tifs-pids/todTIF.htm. Accessed August 23, 2011.

[104] More information about the Dallas TOD TIF District is in Chapter III, Section H.

REFERENCES

Center for Transit-Oriented Development. *Capturing the Value of Transit*. Prepared for FTA. November 2008. http://reconnectingamerica.org/resource-center/books-and-reports/2008/capturing-the-value-of-transit-3.

Center for Transit-Oriented Development. *Transit-Oriented Development Strategic Plan/Metro TOD Program: Detailed Recommendations and Background*. 2011. http://ctod.org/portal/PortlandTOD-Detailed-Recommendations-and-Background.

Center for Transit-Oriented Development. *Transit Revitalization Investment Districts (TRID): Opportunities and Challenges for Implementation*. 2011. http://www.reconnectingamerica.org/news-center/reconnecting-america-news/2011/evaluating-transit-revitalization-investment-districts-in-pennsylvania.

City of Dallas Economic Development. "Transit Oriented Development (TOD) Tax increment Financing (TIF) District." http://www.dallas-ecodev.org/incentives/tifs-pids/todTIF.htm. Accessed August 23, 2011.

Council of Development Finance Agencies. "CDFA Online Resource Database." http://www.cdfa.net/cdfa/cdfaweb.nsf/pages/tifcbuildingresources.html. Accessed August 23, 2011.

E-4.   JOINT DEVELOPMENT[105]

Joint development is generally defined as a real estate development project undertaken by a public agency and a private partner. TOD projects generally involve developing publicly owned land. FTA has guidance on what joint development projects are eligible for public funding under federal transit law.[106] According to this guidance, a joint development project can include "commercial and residential development that is physically or functionally related to public transportation projects; pedestrian and bicycle access to a public transportation facility; construction, renovation, and improvement of intercity bus and intercity rail stations and terminals; and renovation and improvement of historic transportation facilities. Further, to be eligible for federal funding, a joint development project must meet three criteria:

- Enhance economic development or incorporating private investment.

- Enhance the effectiveness of a public transportation project or establish new or enhanced coordination between public transit and other transportation.

- Provide a fair share of revenue to be used for public transit.[107]

Joint development is the only value capture mechanism that is commonly employed directly by transit agencies. It can take many forms, ranging from an agreement to develop land owned by the transit agency to joint financing and development of a larger project that incorporates both transit facilities and private development. A joint development agreement can include a cost-sharing agreement, a revenue-sharing agreement, or a combination of the two. Cost-sharing agreements usually involve cooperation to pay for infrastructure that helps to integrate transit with surrounding development. Revenue-sharing agreements distribute the revenue that results from development among joint development partners. Examples of revenue-sharing agreements include ground-lease revenue, air-rights payments, or in some cases sharing a percentage of rents or other revenue from development. Because many joint development projects are designed to meet multiple goals such as providing affordable housing, local jurisdictions can also help finance aspects of the project.[108]

KEY FACTORS FOR EVALUATION

Applicability to different types of infrastructure: Joint development is most applicable to projects that include development on transit agency-owned land. These types of projects make the most sense where infrastructure improvements (e.g., circulation improvement or structured parking) will benefit all

---

[105] This tool description is based on previous work conducted by the Center for Transit-Oriented Development and Strategic Economics, including the report, *Capturing the Value of Transit*.

[106] DOT. "Notice of Final Agency Guidance on the Eligibility of Joint Development Improvements Under Federal Transit Law." *Federal Register.* Vol. 72, No. 25. February 7, 2007.

[107] Ibid, and Center for Transit-Oriented Development. *Capturing the Value of Transit*. Prepared for FTA. November 2008. p. 26.

[108] Center for Transit-Oriented Development. *Capturing the Value of Transit*. Prepared for FTA. November 2008. p. 26.

parties. Many joint development projects involve building parking structures to replace parking lost through development on existing surface parking lots.

Approval requirements and legal and political considerations: Joint development agreements can take many forms, and the approval requirements vary as described above. There are also specific limitations on the use of land purchased with federal transportation funds.

Application for strong and weak real estate markets: Joint development requires a strong real estate market and a specific development opportunity.

Capacity and scale: The capacity and scale for this tool vary depending on the development opportunity.

Ease of use: Joint development involves complex financial transactions and requires public-sector real estate knowledge and a capable private partner.

USE OF THE TOOL IN PRACTICE

Example: The Highlands at Morristown Train Station

Location: Morristown, New Jersey

Description: In the mid-2000s, New Jersey Transit, the state's public transportation corporation, issued a request for proposals to develop a surface parking lot at Morristown Station on a commuter rail line that provides direct service to Manhattan. The agency chose a developer who proposed to purchase the property and build a five-story, mixed-use building including 228 apartments, 8,000 square feet of ground-floor retail, and a five-story garage with 722 parking spaces for commuters, tenants, and shoppers. The developer paid $7 million of the $8.75 million cost of building the parking garage; New Jersey Transit contributed the remainder and owns and operates 415 commuter spaces in the garage. The agency also receives a portion of the rents generated by the property. The project, known as The Highlands at Morristown Train Station, was completed in 2009.[109]

GETTING STARTED

A transit agency might begin a joint development project by identifying potential development sites and any limitations on the use of the land. Some transit agencies have established real estate offices and implemented real estate and joint development policies.[110]

REFERENCES

Center for Transit-Oriented Development. *Capturing the Value of Transit*. Prepared for FTA. 2008. http://reconnectingamerica.org/resource-center/books-and-reports/2008/capturing-the-value-of-transit-3.

---

[109] Voorhees Transportation Center, Rutgers School of Planning and Public Policy. "Transit Village Update: Morristown Projects Move Forward." *Transit Friendly Development: Newsletter of Transit-Oriented Development and Land Use in New Jersey*. December 2008.

[110] See, for example, Santa Clara Valley Transportation Authority. "Real Estate." http://www.vta.org/realestate. Accessed July 24, 2011.

DOT. "Notice of Final Agency Guidance on the Eligibility of Joint Development Improvements Under Federal Transit Law." *Federal Register.* Vol. 72, No. 25. February 7, 2007. http://edocket.access.gpo.gov/2007/pdf/E7-1979.pdf.

Renne, John L.; Bartholomew, Keith; and Wontor, Patrick. "Transit-Oriented Development: Case Studies and Legal Issues." *Legal Research Digest 36.* August 2011. http://reconnectingamerica.org/assets/Uploads/201108tcrplrd36.pdf.

Voorhees Transportation Center, Rutgers School of Planning and Public Policy. "Transit Village Update: Morristown Projects Move Forward." *Transit Friendly Development: Newsletter of Transit-Oriented Development and Land Use in New Jersey.* December 2008. http://policy.rutgers.edu/vtc/tod/newsletter/vol4-num3/tran_village_update.html#Morristown.

F. GRANTS AND OTHER PHILANTHROPIC SOURCES

F-1.   CONGESTION MITIGATION AND AIR QUALITY IMPROVEMENT PROGRAM

The Congestion Mitigation and Air Quality Improvement (CMAQ) Program funds transportation projects or programs that contribute to improving air quality and relieving congestion. Jointly administered by FHWA and FTA, the CMAQ program was most recently reauthorized in 2012 under Moving Ahead for Progress in the 21st Century (MAP-21).[111] The program provides funding based on a statutory formula to state departments of transportation, metropolitan planning organizations (MPOs), and project sponsors for a growing variety of transportation projects, primarily in nonattainment and maintenance areas.[112]

MPOs typically distribute CMAQ funds, which are available to a wide range of government and nonprofit organizations, as well as private entities contributing to public-private partnerships. MPOs can plan or implement their own air quality programs in addition to approving CMAQ funds for other projects.[113] To pay for TOD infrastructure projects that are not eligible for CMAQ or STP funding, some MPOs exchange CMAQ and STP funds for unrestricted transportation funding from local governments. MPOs can use this same method of exchanging funds to make grants for smaller projects (i.e., below $2 million) that are TOD-related but do not qualify for transportation funding, such as below-ground infrastructure improvements.[114]

KEY FACTORS FOR EVALUATION

Applicability to different types of infrastructure: CMAQ funding can be used for a wide range of capital investments that reduce emissions, including pedestrian and bicycle facilities, public transit improvements, and congestion and traffic flow improvements. To be eligible for CMAQ funds, a project must be included in the MPO's current transportation plan and transportation improvement program (TIP).[115]

Approval requirements and legal and political considerations: Organizations that want access to CMAQ funds must first ask their MPO to include the project in the TIP.

---

[111] FHWA. "Congestion Mitigation and Air Quality Improvement Program (CMAQ)." http://www.fhwa.dot.gov/map21/cmaq.cfm. Accessed September 19, 2012.

[112] Nonattainment areas are those that fail to attain the national ambient air quality standards (NAAQS) for ozone, carbon monoxide, and particulate matter. Maintenance areas are former nonattainment areas with approved plans to maintain NAAQS.

[113] FHWA. "The Congestion Mitigation and Air Quality (CMAQ) Improvement Program: CMAQ Program Assistance, Project Proposals and the Federal Aid Process." http://www.fhwa.dot.gov/environment/air_quality/cmaq/reference/brochure/brochure08.cfm. Accessed August 24, 2011.

[114] Center for Transit-Oriented Development. *Financing Transit-Oriented Development in the San Francisco Bay Area: Policy Options and Strategies.* Prepared for MTC. August 2008.

[115] FHWA. "Congestion Mitigation and Air Quality Improvement Program (CMAQ)." http://www.fhwa.dot.gov/map21/cmaq.cfm. Accessed September 19, 2012.

Application for strong and weak real estate markets: This funding source is applicable to both strong and weak real estate markets because it doesn't depend on new development occurring.

Capacity and scale: The scale of projects that can be funded depends on the size of a state or region's allocation of CMAQ funds as well as each region's priorities for using those funds.

Ease of use: Federal funding sources such as CMAQ have significant administrative and reporting requirements.

Although local match requirements for CMAQ vary by state, generally the federal government will pay for up to 80 percent of eligible project costs; the remainder is the responsibility of the project sponsor. Because of this local match requirement and the scarcity of federal funding for these types of transportation improvements compared to the demand, project sponsors often end up combining many sources of funding for a single project, extending project timelines and creating logistical complications because of the need to accommodate differing funding schedules.

OTHER LIMITATIONS OF THE TOOL

CMAQ funds cannot be used for certain types of infrastructure that are critical for TOD, such as land assembly or building parking facilities. To fund these types of infrastructure, several MPOs, including the San Francisco Bay Area MTC, Portland Metro, and the North Central Texas Council of Governments, trade federal transportation funds passed through to the regions (including CMAQ as well as STP and Urbanized Area Formula funding, discussed in Section F-3 of this appendix) for local unrestricted sources that transit agencies or municipalities have allocated to eligible uses, such as road maintenance. [116] While trading federal funds allows regions to allocate their funds more flexibly, the practice also creates uncertainty and can delay projects.

USE OF THE TOOL IN PRACTICE

Example: Bike Station and BikeLink Parking Technology at Transit Stations

Location: Folsom, California

Description: The Sacramento Area Council of Governments awarded Folsom, California $158,000 of CMAQ funds through their 2008 Regional Bicycle & Pedestrian Funding Program to build a new bike station and retrofit 22 bike lockers with BikeLink technology and purchase an additional 50 BikeLink lockers. [117,118] BikeLink cards allow bicyclists to have around-the-clock access to secure bicycle lockers. The project improves bicycle access to two light rail stations as well as to a major transfer point for local bus service. The city of Folsom provided matching funds of $18,000.

---

[116] For smaller projects (i.e., less than $2 million), this exchange also removes federal environmental review and labor requirements that would outweigh the value of the grant.

[117] Sacramento Area Council of Governments. "Bicycle and Pedestrian Funding Program—List of Approved Projects." April 17, 2008.

[118] City of Folsom, Parks & Recreation Department. "Application for Funding under the SACOG Bicycle and Pedestrian Funding Program." December 3, 2007.

GETTING STARTED

Local governments should contact their MPOs for information about the process for placing projects on the TIP and with suggestions for CMAQ projects.

REFERENCES

Center for Transit-Oriented Development. *Financing Transit-Oriented Development: Framing the Issues and Assessing the Tools.* Prepared for Transportation for America (unpublished). June 2011.

Center for Transit-Oriented Development. *Financing Transit-Oriented Development in the San Francisco Bay Area: Policy Options and Strategies.* Prepared for MTC. August 2008. http://www.mtc.ca.gov/planning/smart_growth/tod/Financing_TOD_in_SFBA.pdf.

City of Folsom, Parks & Recreation Department. "Application for Funding under the SACOG Bicycle and Pedestrian Funding Program." December 3, 2007. http://www.sacog.org/regionalfunding/fundingprograms/pdf/2008/winners/City%20of%20Folsom%20Bikelink.pdf.

FHWA. "Congestion Mitigation and Air Quality Improvement Program (CMAQ)." http://www.fhwa.dot.gov/map21/cmaq.cfm. Accessed September 19, 2012.

FHWA. "The Congestion Mitigation and Air Quality (CMAQ) Improvement Program: CMAQ Program Assistance, Project Proposals and the Federal Aid Process." http://www.fhwa.dot.gov/environment/air_quality/cmaq/reference/brochure/brochure08.cfm. Accessed August 24, 2011.

FHWA. *The Congestion Mitigation and Air Quality Improvement (CMAQ) Program under the Safe, Accountable, Flexible, Efficient Transportation Equity Act: A Legacy for Users: Final Program Guidance.* 2008. http://www.fhwa.dot.gov/environment/air_quality/cmaq.

Sacramento Area Council of Governments. "Bicycle and Pedestrian Funding Program—List of Approved Projects." April 17, 2008. http://www.sacog.org/calendar/2008/04/24/bikeped/pdf/04-List%20of%20approved%20proj.%27s.pdf.

F-2. TRANSPORTATION ALTERNATIVES PROGRAM (FORMERLY TRANSPORTATION ENHANCEMENTS PROGRAM)

The Transportation Alternatives Program provides funding to expand and improve transportation options. It is funded from each state's federal apportionment and is administered by the states. It was most recently authorized in 2012 under MAP-21. The program was previously known as the Transportation Enhancements Program.

State DOTs and MPOs must establish a competitive process to award funding. The entities eligible for funding include local governments, regional transportation authorities, transit agencies, natural resource or public land agencies, and tribal governments. [119]

The amount of federal transportation funding designated for transportation enhancements could not meet the need for station-area infrastructure and other public improvements, and MAP-21 further reduced funding while also allowing states more flexibility to redirect transportation alternatives money to other uses.

KEY FACTORS FOR EVALUATION

Applicability to different types of infrastructure: This tool can be applied to construction, planning, and design of bicycle and pedestrian facilities including sidewalks, pedestrian and bicycle signals, traffic calming techniques, lighting, and safety-related infrastructure, among other things. [120]

Approval requirements and legal and political considerations: The amount states receive for transportation alternatives is 2 percent of the amount awarded to nine different transportation programs. In fiscal year 2013, the amount is $808,760,000. [121]

Application for strong and weak real estate markets: This funding source is applicable to both strong and weak real estate markets because it does not depend on new development occurring.

Capacity and scale: The scale of projects that can be funded with transportation alternatives varies depending on a region's allocation of transportation alternatives funding and priorities for using that funding.

Ease of use: Federal funding sources such as transportation alternatives have significant administrative and reporting requirements.

---

[119] National Transportation Enhancements Clearinghouse. "Memorandum: MAP-21 and Its Effects on Transportation Enhancements." July 13, 2012.

[120] FHWA. "Moving Ahead for the 21st Century Act (MAP-21): Transportation Alternatives Definitions." http://www.fhwa.dot.gov/environment/transportation_enhancements/legislation/map21.cfm. Accessed September 20, 2012.

[121] National Transportation Enhancements Clearinghouse. "MAP-21 and Transportation Alternatives Apportionments." July 17, 2012.

The federal government will pay for up to 80 percent of a transportation alternatives project; the remainder is the responsibility of the project sponsor.[122] Because of this local match requirement and the scarcity of federal funding for these types of transportation improvements compared to the demand, transportation alternatives project sponsors often end up combining many sources of funding for a single project, extending project timelines and creating logistical challenges involving the need to accommodate different funding schedules.

OTHER LIMITATIONS OF THE TOOL

Similar to CMAQ funding, transportation alternatives funding cannot be used for certain types of infrastructure that are critical for TOD, such as land assembly or building parking facilities, and some MPOs have traded these federal funds for local funds, as described in Chapter III, Section I. These trades allow regions to use their funds more flexibly but also can create uncertainty and delays.

USE OF THE TOOL IN PRACTICE

Example: 28th Street Improvements Project

Location: Boulder, Colorado (at the planned Boulder Transit Village)

Description: The 28th Street Improvements Project transformed 2 ½ miles of 28th Street into an attractive, multimodal corridor by adding new regional bus service and making extensive pedestrian and bicycle improvements. The corridor serves as a gateway for the city of Boulder and the planned Boulder Transit Village, which will be built around a stop on the planned U.S. 36 Bus Rapid Transit Corridor. Part of the funding came from the Transportation Enhancements program, including $395,000 for a pedestrian crossing (with a local match of $395,000) and $600,000 for bicycle facilities (with a local match of $150,000).[123]

GETTING STARTED

State departments of transportation are responsible for developing and administering their own Transportation Alternatives Programs, so the process of applying for funding varies by state. In general, for previous Transportation Enhancements funding local governments and nonprofits partnered with governments apply to their state department of transportation.

REFERENCES

Center for Transit-Oriented Development. *Financing Transit-Oriented Development: Framing the Issues and Assessing the Tools.* Prepared for Transportation for America (unpublished). June 2011.

---

[122] National Transportation Enhancements Clearinghouse. *Transportation Alternatives: Program Manual Development.* 2012.

[123] City of Boulder. "28th Street Improvements Project: Project Overview." http://www.bouldercolorado.gov/index.php?option=com_content&task=view&id=294&Itemid=1198. Accessed August 30, 2011.

Center for Transit-Oriented Development. *Financing Transit-Oriented Development in the San Francisco Bay Area: Policy Options and Strategies.* Prepared for MTC. 2008. http://www.mtc.ca.gov/planning/smart_growth/tod/Financing_TOD_in_SFBA.pdf.

City of Boulder. "28th Street Improvements Project: Project Overview." http://www.bouldercolorado.gov/index.php?option=com_content&task=view&id=294&Itemid=1198. Accessed August 30, 2011.

FHWA. "Moving Ahead for the 21st Century Act (MAP-21): Transportation Alternatives Definitions." http://www.fhwa.dot.gov/environment/transportation_enhancements/legislation/map21.cfm. Accessed September 20, 2012.

National Transportation Enhancements Clearinghouse. "MAP-21 and Transportation Alternatives Apportionments." July 17, 2012. http://www.enhancements.org/download/Publications/Briefs/MAP-21_Apportionments.pdf.

National Transportation Enhancements Clearinghouse. "Memorandum: MAP-21 and Its Effects on Transportation Enhancements." July 13, 2012. http://www.enhancements.org/download/Publications/Briefs/MAP-21_and_Transportation_Enhancements.pdf.

National Transportation Enhancements Clearinghouse. *Transportation Alternatives: Program Manual Development.* 2012. http://www.enhancements.org/download/Publications/Briefs/Program_Manual_Development.pdf.

F-3.   URBANIZED AREA FORMULA FUNDING PROGRAM

The Urbanized Area Formula Funding Program, most recently authorized in 2012 under MAP-21, makes federal resources available to urbanized areas[124] and to state governors for transit capital, operating assistance, and transportation-related planning. MAP-21 eliminated a requirement that 1 percent of each urbanized area's apportionment be used for transit enhancements, which included pedestrian and bicycle access improvements, transit connections to parks, and enhanced access for persons with disabilities.[125]

For urbanized areas with populations of 200,000 or more, Urbanized Area Formula funds flow directly to a designated recipient selected locally to apply for and receive federal funds—typically the region's MPO. For urbanized areas under 200,000 in population, the funds are apportioned to each state for distribution. A few areas under 200,000 in population have been designated as transportation management areas and receive apportionments directly. In fiscal year 2012, FTA allocated approximately $2.28 billion to the Urbanized Area Formula program.[126]

KEY FACTORS FOR EVALUATION

Applicability to different types of infrastructure: Urbanized Area Formula funds can be used for capital projects, planning, and job access and reverse commute projects. Operating costs for public transportation are also eligible for funding in areas with fewer than 200,000 people. The definition for a capital project includes a project for a joint development improvement that may include "property acquisition; demolition of existing structures; site preparation; utilities; building foundations; walkways; pedestrian and bicycle access to a public transportation facility; construction, renovation, and improvement of intercity bus and intercity rail stations and terminals; renovation and improvement of historic transportation facilities; open space; safety and security equipment and facilities...; a capital project for, and improving, equipment or a facility for an intermodal transfer facility or transportation mall; and construction of space for commercial uses."[127]

Approval requirements and legal and political considerations: Similar to the CMAQ funds, organizations that want access to Urbanized Area Formula funds work with their MPO to place the project on the TIP.

Application for strong and weak real estate markets: This funding source is applicable to both strong and weak real estate markets because it does not depend on new development occurring.

Capacity and scale: The scale of projects that can be funded with Urbanized Area Formula varies depending on a region's funding allocation and priorities.

Ease of use: Federal funding sources such as the Urbanized Area Formula have significant administrative and reporting requirements.

---

[124] An urbanized area is an incorporated area with a population of 50,000 or more designated by the U.S. Census.

[125] FTA. "Chapter 53 of title 49, United States Code, as amended by MAP-21." 2012.

[126] FTA. "FY 2012 Section 5307 and Section 5340 Urbanized Area Apportionments." http://www.fta.dot.gov/printer_friendly/12853_14254.html. Accessed August 2011.

[127] FTA. "Chapter 53 of title 49, United States Code, as amended by MAP-21." 2012.

OTHER LIMITATIONS OF THE TOOL

In general, the federal share of a project funded with Urbanized Area Formula cannot exceed 80 percent of the project cost.

USE OF THE TOOL IN PRACTICE

Example: Union Station Interconnect

Location: New Haven, Connecticut

Description: This 2008 project expanded bicycle access to Union Station, the intermodal center of Connecticut passenger rail service.[128] The project included the development of 4.6 miles of bicycle lanes and associated road safety improvements to connect the station to downtown and other neighborhoods. It also included 100 new bicycle parking spaces at the station. Funding included $130,500 in transit enhancement funds under the Urbanized Area Formula Program and a $14,500 local match.[129]

GETTING STARTED

To access Urbanized Area program funds, local governments typically work with their MPOs to place projects on the region's TIP.

REFERENCES

City of New Haven. *New Haven Union Station Transit-Oriented Development Study.* February 2008. http://www.cityofnewhaven.com/CityPlan/pdfs/UnionStationTOD.pdf.

FHWA. "Transportation Enhancements Administered by the Federal Transit Administration." http://www.fhwa.dot.gov/environment/transportation_enhancements/guidance/te_provision.cfm. Accessed August 30, 2011.

FTA. "FY 2012 Section 5307 and Section 5340 Urbanized Area Apportionments." http://www.fta.dot.gov/printer_friendly/12853_14254.html. Accessed August 2011.

FTA. "Chapter 53 of title 49, United States Code, as amended by MAP-21." 2012. http://www.fta.dot.gov/documents/chapter53redlineMAP21.pdf.

South Central Regional Council of Governments. "2008 Transit Enhancement Project Proposals." August 13, 2008. http://www.scrcog.org/documents/Transit_Enhancement%20Rpt2008.pdf.

---

[128] City of New Haven. *New Haven Union Station Transit-Oriented Development Study.* February 2008.

[129] South Central Regional Council of Governments. "2008 Transit Enhancement Project Proposals." August 13, 2008.

F-4.   COMMUNITY DEVELOPMENT BLOCK GRANT PROGRAM

The Community Development Block Grant (CDBG) Program, administered by HUD, supports the provision of decent, affordable housing and community services, job creation, and retention of businesses in vulnerable neighborhoods. The CDBG program provides annual grants on a formula basis to 1,209 local government agencies and states in the following program areas:

- **Entitlement Communities Grants:** This program allocates annual grants to larger cities and urban counties to provide housing and expand economic opportunities, principally for low- and moderate-income people.

- **State-Administered CDBG (also known as the Small Cities CDBG program):** This program provides money to each state to award grants to smaller cities and counties for community development activities.

- **Section 108 Loan Guarantee Program:** This program provides financing for economic development; housing rehabilitation; and public facilities rehabilitation or construction for low- to moderate-income people.

- **Neighborhood Stabilization Program:** This program provides grants to communities hardest hit by foreclosures and delinquencies to purchase, rehabilitate, or redevelop homes and stabilize neighborhoods.

- **Renewal Communities/Empowerment Zones/Enterprise Communities:** This program aims to attract private investment for sustainable economic and community development.[130]

While most of these funding programs might not be applicable to TOD infrastructure as defined in this report, they can make TOD infrastructure possible by funding a portion of a larger TOD project that includes an infrastructure component. The Section 108 Loan Guarantee Program is the most relevant to providing TOD infrastructure. The program allows communities to leverage CDBG funds into federally guaranteed loans for economic revitalization projects. Eligible uses for Section 108 financing include:

- Economic development activities eligible under CDBG.

- Acquisition of real property.

- Rehabilitation of publicly owned real property.

- Housing rehabilitation eligible under CDBG.

- Construction, reconstruction, or installation of public facilities (including street, sidewalk, and other site improvements).

- Related relocation, clearance, and site improvements.

---

[130] HUD. "Community Development Block Grant Program—CDBG." http://portal.hud.gov/hudportal/HUD?src=/program_offices/comm_planning/communitydevelopment/programs. Accessed August 31, 2011.

- Payment of interest on the guaranteed loan and issuance costs of public offerings.

- Debt service reserves.

- Housing construction (in limited circumstances).[131]

KEY FACTORS FOR EVALUATION

Applicability to different types of infrastructure: This funding could be used for property acquisition, rehabilitation of publicly owned real property, streets, and sidewalks. CDBG funding could also be used in combination with other funding and financing tools to contribute to a larger TOD project that meets the criteria for using CDBG funds.

Approval requirements and legal and political considerations: HUD determines grant amounts with formulas based on the extent of poverty, population, housing overcrowding, age of housing, and population growth lag in relationship to other metropolitan areas. Most larger cities and urban counties receive annual CDBG allocations; smaller communities can access CDBG funding through state agencies. At least 70 percent of CDBG funds must be used for activities that benefit low- and moderate-income people.[132]

Application for strong and weak real estate markets: This funding source is designed to benefit low- and moderate-income people, prevent or eliminate slums or blight, and address urgent community development needs. Therefore projects are usually located in areas that have weak real estate markets.

Capacity and scale: The size of a project that CDBG can fund depends on a city or county's allocation of and priorities for CDBG funding. As described above, CDBG funds are apportioned to communities based on HUD formulas. Section 108 Loan Guarantees allow a community to leverage a portion of CDBG grant money to finance a larger loan.

Ease of use: Federal funding sources such as CDBG have significant administrative and reporting requirements.

OTHER LIMITATIONS OF THE TOOL

There are significant restrictions on how CDBG funds can be spent. CDBG funds are not applicable to TOD infrastructure projects as defined in this report unless the infrastructure projects are part of a larger TOD project that meets the criteria of CDBG, such as an affordable housing or economic revitalization project.

---

[131] HUD. "Section 108 Loan Guarantee Program," http://portal.hud.gov/hudportal/HUD?src=/program_offices/comm_planning/communitydevelopment/programs/108. Accessed August 31, 2011.

[132] HUD. "Community Development Block Grant Program—CDBG." http://portal.hud.gov/hudportal/HUD?src=/program_offices/comm_planning/communitydevelopment/programs. Accessed August 31, 2011.

USE OF THE TOOL IN PRACTICE

CDBG funds have been used as part of the funding and financing package for many TOD projects.

Example: EastLake Streetscape Improvement Project

Location: Oakland, California (AC Transit and Lake Merritt BART Station)

Description: The city of Oakland widened sidewalks and repainted crosswalks, added bulb-outs at intersections and bus stops, and installed pedestrian amenities in the transit-accessible neighborhood of EastLake. Funding sources included $85,000 from CDBG, $1,730,000 in MTC grants, $200,000 from the Transportation Fund for Clean Air, $200,000 from the Oakland Capital Improvement Program, $442,000 from a transportation sales tax (Measure B), and $412,000 in other local funds.

GETTING STARTED

CDBG entitlement communities are eligible to apply for assistance through the section 108 loan guarantee program. CDBG non-entitlement communities can also apply, provided their state agrees to pledge the CDBG funds necessary to secure the loan. Applicants can receive a loan guarantee directly or designate another public entity, such as an industrial development authority, to carry out their Section 108-assisted project.

REFERENCES

MTC. *Ten Years of TLC: An Evaluation of MTC's Transportation for Livable Communities. Appendix A: Case Studies.* 2007. http://www.mtc.ca.gov/planning/smart_growth/tlc/tlc_eval/TLC_Evaluation_App_A_Case_Studies.pdf.

HUD. "Community Development Block Grant Program— CDBG." http://portal.hud.gov/hudportal/HUD?src=/program_offices/comm_planning/communitydevelopment/programs. Accessed August 31, 2011.

HUD. "Facts About Farmworkers and Colonias." http://www.hud.gov/groups/farmwkercolonia.cfm. Accessed August 31, 2011.

HUD. "Section 108 Loan Guarantee Program." http://portal.hud.gov/hudportal/HUD?src=/program_offices/comm_planning/communitydevelopment/programs/108. Accessed August 31, 2011.

F-5.   ECONOMIC DEVELOPMENT ADMINISTRATION GRANTS

The Economic Development Administration (EDA), an agency in the U.S. Department of Commerce, provides grants to economically distressed communities to generate new employment, help retain existing jobs, and stimulate industrial and commercial growth. EDA recognizes TOD as an economic development tool and has provided funding for TOD projects.[133] EDA investment programs include:

- **Public Works and Economic Development Assistance:** Empowers distressed communities to revitalize, expand, and upgrade their physical infrastructure to attract new industry, encourage business expansion, diversify local economies, and generate or retain long-term, private-sector jobs and investment.[134] The investments are also designed to help communities attract private capital investment and higher-skill, higher-wage jobs. The funds can be used for:

  o   Acquisition or development of land and improvements for use in a public works, public service, or other type of development facility; or

  o   Acquisition, design and engineering, construction, rehabilitation, alteration, expansion, or improvement of such a facility, including related machinery and equipment.[135]

- **Economic Adjustment Assistance:** Helps state and local interests design and implement strategies to adjust or bring about change to an economy. The program focuses on areas that have experienced or are under threat of serious damage to the underlying economic base.[136]

- **Partnership Planning:** Supports local organizations (e.g., economic development districts and Indian tribes) with long-term planning efforts.[137]

- **Global Climate Change Mitigation Incentive Fund:** Finances projects that foster economic development by advancing the green economy in distressed communities.[138]

KEY FACTORS FOR EVALUATION

Applicability to different types of infrastructure: EDA funding can be spent on a variety of infrastructure types provided the project has an economic development purpose.

Approval requirements and legal and political considerations: EDA has an application process and approves grants on a quarterly basis.

---

[133] American Planning Association. "EDA Update." July 2009.

[134] EDA. "EDA Investment Programs." http://www.eda.gov/programs.htm. Accessed August 24, 2012.

[135] U.S. Office of Management and Budget. *A-133 Compliance Supplement: Department of Commerce.* 2011. p. 4-11.300-3.

[136] EDA. "EDA Investment Programs." http://www.eda.gov/programs.htm. Accessed August 24, 2012.

[137] Ibid.

[138] FedCenter.gov. "Global Climate Change Mitigation Incentive Fund." http://www.fedcenter.gov/Bookmarks/index.cfm?id=13420&pge_id=1854. Accessed August 24, 2012.

Application for strong and weak real estate markets: EDA funds are available only to communities meeting the agency's criteria for economic distress, so EDA-funded projects are usually located in areas with weak real estate markets.

Capacity and scale: There are no minimum or maximum investment amounts for an EDA grant. EDA can fund up to 80 percent of the total project cost if a community meets the criteria.

Ease of use: Federal funding sources such as EDA grants have significant administrative and reporting requirements.

USE OF THE TOOL IN PRACTICE

EDA funds are applicable to a wide variety of infrastructure project types, provided the community and project meet the program criteria.

Example: West Broadway Urban Village Infrastructure Grant

Location: Seaside, California

Description: EDA provided grant funding for the city of Seaside to take the first step toward implementing the West Broadway Urban Village Specific Plan, which envisioned the West Broadway area as the core of a new pedestrian- and bicycle-friendly, transit-oriented urban village. The infrastructure improvement phase of the project includes streetscape and intersection improvements, development of pedestrian and bicycle amenities, and the upgrade of public utilities within the public right of way. Funding sources for the $1.35 million project included a $945,000 EDA grant and $400,000 in local matching CDBG funds.[139]

GETTING STARTED

EDA recommends that all potential applicants contact the appropriate Economic Development Representative or point of contact for their state. EDA also holds training sessions for potential applicants.

REFERENCES

American Planning Association. "EDA Update." July 2009. http://www.planning.org/eda/newsletter/2009/jul.htm.

City of Seaside. "Current Projects: West Broadway Urban Village Infrastructure Grant." http://ci.seaside.ca.us/index.aspx?page=381. Accessed August 31, 2011.

EDA. "EDA Investment Programs." http://www.eda.gov/programs.htm. Accessed August 24, 2012

EDA. "Federal Funding Opportunities." http://www.eda.gov/ffo.htm. Accessed July 26, 2012.

---

[139] City of Seaside. "Current Projects: West Broadway Urban Village Infrastructure Grant." http://ci.seaside.ca.us/index.aspx?page=381. Accessed August 31, 2011.

EDA. "New Investments: California." http://www.eda.gov/NewsEvents/NewInvestments/ca.xml. Accessed August 31, 2011.

FedCenter.gov. "Global Climate Change Mitigation Incentive Fund." http://www.fedcenter.gov/Bookmarks/index.cfm?id=13420&pge_id=1854. Accessed August 24, 2012.

U.S. Office of Management and Budget. *A-133 Compliance Supplement: Department of Commerce.* 2011. http://www.whitehouse.gov/sites/default/files/omb/assets/OMB/circulars/a133_compliance/2011/doc.pdf.

F-6. FOUNDATION GRANTS

Foundations, including private foundations and public charities, are nongovernmental organizations that make grants with a charitable purpose. Foundations make many types of grants, including:[140]

- Capital or capital campaign: A capital grant provides funds to purchase property or equipment, build a facility, or remodel or expand a facility.

- Operating or general support: An operating grant provides support for the day-to-day costs of running an organization.

- Endowment: An endowment fund is a permanent annual source of income for an organization's operating or project expenses.

- Unrestricted: An unrestricted grant allows the organization to use funds where it needs them most.

- Project: A project grant supports a specific activity.

- Seed: A seed grant helps to jump-start a new organization or a new project or launch a capital campaign.

- Challenge or matching: A challenge or matching grant helps a nonprofit organization leverage additional dollars through a fundraising campaign.

- Pledge: A pledge is a promise to pay in the future.

Foundations are interested in providing support to TOD through traditional and non-traditional means, including grants, technical assistance, and program-related investments (see the next tool profile for more information on program-related investments).[141] However, while foundations have been showing increasing interest in TOD activities, most of their efforts have been related to providing affordable housing or social services around transit facilities or even the transit itself, not the infrastructure needed to support TOD.

A few foundations have provided grants for land or property acquisition and environmental assessment or remediation for brownfield redevelopment. However, most grants for TOD projects have been for pre-development activities, including planning, technical assistance, and community engagement. These pre-development grants are typically made to nonprofit organizations, not to government agencies.

---

[140] Foundation Center. "Grants Classification." http://foundationcenter.org/gainknowledge/grantsclass/ntee_gcs.html. Accessed August 2011.

[141] Katherine Pease & Associates. "Convening on Transit Oriented Development: The Foundation Perspective." Prepared for Center for Transit-Oriented Development, Living Cities, and Boston College Institute for Responsible Investments. February 2009.

KEY FACTORS FOR EVALUATION

Applicability to different types of infrastructure: Foundations have shown the most interest in pre-development activities associated with equitable TOD, such as mixed-income TOD projects. Foundations have also shown interest in establishing TOD land acquisition funds.

Approval requirements and legal and political considerations: Using foundation funding does not require voter approval. Grants are typically allocated through a competitive application process.

Application for strong and weak real estate markets: Depending on the mission and policies of the foundation, grants can be available in strong and weak real estate markets.

Capacity and scale: Grant sizes vary by foundation and grant program.

Ease of use: Foundation grants typically have significant reporting requirements to ensure that the grants are used in compliance with foundation goals.

OTHER LIMITATIONS OF THE TOOL

Because grant programs are typically competitive, grants are unpredictable funding sources.

USE OF THE TOOL IN PRACTICE

This tool is largely untested for TOD infrastructure, but foundations are increasingly funding planning, technical assistance, and community engagement activities related to TOD, as well as land acquisition and brownfield remediation.

Example: Living Cities Integration Initiative

Location: Baltimore; Cleveland; Detroit; Newark, New Jersey; and Minneapolis-St. Paul

Description: The Living Cities Integration Initiative[142] provided $85 million in grants and loans to five cities (Baltimore, Cleveland, Detroit, Newark, and the Twin Cities) for initiatives that encourage public, private, nonprofit, and philanthropic sectors to work together to make communities work for low-income people. Projects in Baltimore and Minneapolis-St. Paul have a focus on TOD. The Minneapolis-St. Paul project will convene local, regional, and state government; the private sector; and nonprofit and philanthropic organizations to create and preserve transit-accessible affordable housing and mixed-use, mixed-income developments; help small businesses deal with disruptions caused by transit corridor construction; and catalyze neighborhood-led development along three regional transit lines. The Baltimore Integration Project[143] will focus on "creating job opportunities and improving neighborhoods

---

[142] Living Cities. "The Integration Initiative." http://www.livingcities.org/integration/. Accessed August 31, 2011.

[143] Baltimore Integration Partnership. "What is the Baltimore Integration Partnership?" http://www.abagrantmakers.org/page/BaltimorePartnership/. Accessed August 31, 2011.

in Central and East Baltimore, while preparing residents for opportunities created by the construction of the Red Line, a 14-mile east-west transit line."[144]

GETTING STARTED

Most grants require a competitive application process. Successful grant-seekers often have established relationships or partnerships with foundations before applying for grants. Many online resources, such as the Foundation Center, provide information on foundations and grants and on writing proposals and grant requests.

REFERENCES

Baltimore Integration Partnership. "What is the Baltimore Integration Partnership?" http://www.abagrantmakers.org/page/BaltimorePartnership/. Accessed August 31, 2011.

Foundation Center. "Grants Classification." http://foundationcenter.org/gainknowledge/grantsclass/ntee_gcs.html. Accessed August 2011.

Katherine Pease & Associates. "Convening on Transit Oriented Development: The Foundation Perspective." Prepared for Center for Transit-Oriented Development, Living Cities, and Boston College Institute for Responsible Investments. February 2009. http://www.katherinepease.com/Convening%20on%20TOD%20-%20The%20Foundation%20Perspective.pdf.

Living Cities. "The Integration Initiative." http://www.livingcities.org/integration/. Accessed August 31, 2011.

Living Cities Integration Initiative. "National Collaborative Announces $80 Million Investment in Five Cities." October 28, 2010. http://backend.livingcities.org/_backend.livingcities.org/files/TII_National_pressrelease.pdf.

---

[144] Living Cities Integration Initiative. "National Collaborative Announces $80 Million Investment in Five Cities." October 28, 2010. http://backend.livingcities.org/_backend.livingcities.org/files/TII_National_pressrelease.pdf, p. 3.

F-7.   PROGRAM-RELATED INVESTMENTS

Program-related investments (PRIs) are mission-driven investments made by foundations to support their philanthropic goals and leverage their funds. Unlike grants, PRIs involve a potential return on the investment. PRIs can include loans, loan guarantees, and equity investments, although they typically come in the form of a low-interest loan. For the recipient, the primary benefit of a PRI is access to capital at lower rates than might otherwise be available. For the funder, the principal benefit is that the repayment or return of equity can be recycled for another charitable purpose, assuming the investment is repaid.

Not all foundations in the United States make PRIs, and those that do typically do so as a supplement to their existing grant programs. PRIs are available only if an applicant has the potential to generate income to repay a loan. PRIs could provide financing for projects that were unable to secure financing from traditional sources. While a large portion of PRI funding in the past has supported affordable housing and community development, in some cases these investments have funded capital projects ranging from rehabilitating historic buildings to preserving open space and wildlife habitat.

The Internal Revenue Service requires interest rates to be below market rates for an investment to qualify as a PRI. However, rates for PRIs can vary depending on the level of risk involved. PRIs can be either secured (i.e., guaranteed by collateral) or unsecured. Some are structured so that a portion of the principal plus interest is paid in regular installments over a set period of time, while others are set up with interest-only payment schedules with a larger payment at the end to cover the principal. The duration of PRIs vary; although most are fairly short-term, PRIs have been used to support multiyear community development projects requiring long-term, "patient" capital.

KEY FACTORS FOR EVALUATION

Applicability to different types of infrastructure: Because foundations require repayment for PRIs, this tool would typically apply only to revenue-generating infrastructure. However, PRIs could be repaid with another revenue source.

Approval requirements and legal and political considerations: This tool does not require voter approval. However, because PRIs involve loan documents and other financial agreements, they typically require more legal expertise than grants. [145]

Application for strong and weak real estate markets: Depending on the mission and policies of the foundation, PRIs can be applicable in strong and weak real estate markets.

Capacity and scale: PRIs range from as little as $1,000 to several million dollars. Generally, the amount depends on the recipient's need and capacity as well as the foundation's scope, size, and risk tolerance. [146]

---

[145] Benabentos, L. et al. *Strategies to Maximize Your Philanthropic Capital: A Guide to Program Related Investments.* Mission Investors Exchange, the Thomson Reuters Foundation, and Linklaters LLP. 2012.
[146] Ibid.

Ease of use: PRIs range from straightforward short-term loans to complex financial transactions. Borrowers administer PRIs like other loans, except that foundations might have additional reporting requirements to ensure that the funds are used in compliance with foundation goals. From a foundation's perspective, PRIs can be administratively complex and are often managed with help from outside consultants.

USE OF THE TOOL IN PRACTICE

A 2009 report on foundation support for TOD found that "for the most part, funders have not made significant PRIs for TOD-specific investments, although there is considerable interest in moving in that direction. PRIs in this arena primarily have been used to support TOD-related property acquisition efforts, including paying for upfront support for a land acquisition fund, or, supporting a citywide land acquisition fund for affordable housing."[147]

Example: Living Cities Catalyst Fund

Location: United States

Description: In 2008, seven foundations invested a total of more than $20 million in the Living Cities Catalyst Fund.[148] The Catalyst Fund provides below-market loans and guarantees to nonprofit organizations that create opportunities and make markets work for low-income communities. Most investments have not been focused on TOD. However, in 2011, the fund invested $3 million in the Bay Area Transit Oriented Affordable Housing Fund.[149]

GETTING STARTED

Project sponsors interested in PRIs should approach individual foundations.

REFERENCES:

Benabentos, Lucia; Storms, Justin; Teuscher, Carlos; and Van Loo, Jon. *Strategies to Maximize Your Philanthropic Capital: A Guide to Program Related Investments.* Mission Investors Exchange, the Thomson Reuters Foundation, and Linklaters LLP. 2012. http://www.missioninvestors.org/tools/strategies-maximize-your-philanthropic-capital-guide-program-related-investments-primer.

Foundation Center. "Home page." http://foundationcenter.org. Accessed August 31, 2011.

Katherine Pease & Associates. "Convening on Transit Oriented Development: The Foundation Perspective." Prepared for Center for Transit-Oriented Development, Living Cities, and Boston College Institute for Responsible

---

[147] Katherine Pease & Associates. "Convening on Transit Oriented Development: The Foundation Perspective." Prepared for Center for Transit-Oriented Development, Living Cities, and Boston College Institute for Responsible Investments. February 2009. p 3.

[148] Living Cities. "Catalyst Fund." http://www.livingcities.org/investment/vehicles/catalyst-fund. Accessed August 31, 2011.

[149] More information about the fund is in Chapter III, Section J.

Investments. February 2009. http://www.katherinepease.com/Convening%20on%20TOD%20-%20The%20Foundation%20Perspective.pdf.

Living Cities. "Catalyst Fund." http://www.livingcities.org/investment/vehicles/catalyst-fund. Accessed August 31, 2011.

G. EMERGING TOOLS

G-1. STRUCTURED FUNDS

A structured fund is a pooled equity that is structured to be relatively low risk with opportunity for limited returns. Structured funds invest in two types of assets: fixed-income products, like mortgages or mortgage-backed securities where the income stream is fixed, and derivatives, which are more risky but offer more opportunity for gains from upward movement in markets. The term for these funds can vary from three to 10 years, and they are tied to a specific source of capital production.

However, the term "structured funds" has also begun to take on a secondary meaning in community development, where the risk profile for the capital source is typically higher than for a conventional fund and the term of the fund might need to be longer. In this context, structured funds typically pool contributions from multiple investors with different risk and return parameters. The different risk profiles allow the fund's manager to blend interest rates and provide lower-cost, limited-recourse loans with higher loan-to-value ratios. Capital from each investor is placed into a tiered structure, or "capital stack," based on risk profile, so that the investors willing to accept more risk are in the stack above other investors with less risk tolerance. Thus each layer in the stack "protects" the next layer down from risk. This tiered structure mitigates risk to maximize the leverage from the funding sources.

Funds with a social mission vary in their goals, activities, and sources of capital. Mission-driven structured funds often attract grant funding from public-sector entities that can be dispersed without return expectations, thereby allowing them to occupy the critical top risk absorption position and leverage other investment with lower risk tolerance (top-loss position in the capital stack). Other investors can include foundations using program-related investment funds that have below-market rate return expectations, community development finance institutions that make below-market rate loans, and commercial banks that can use these funds as part of their obligation under the Community Reinvestment Act to help meet the credit needs of the communities in which they operate. Borrowers make payments back into the fund, which can revolve to allow additional lending or is held as security until the fund expires and investors are repaid.[150] The funds are typically "closed ended," meaning they do not revolve indefinitely. Community development-oriented structured funds are typically targeted to land and property acquisition that support affordable housing because there is a clear source of "capital production," or take-out funding, associated with funding these projects. The loans are used to acquire land and/or property before all of the project funding is in place, but once the project has its funding, the acquisition loan is paid off. In most cases, the acquisition funds also carry a longer term (e.g., five to 10 years) than a conventional loan or even than funding from a conventional structured fund. Loans can have a 110-percent loan-to-value ratio so that some of the loan money can be used for predevelopment activities.

Nationally, approximately 15 affordable housing-related structured loan funds are operating or under development, including three that are dedicated to transit-oriented locations:

---

[150] Center for Transit-Oriented Development and Strategic Economics. *San Francisco Bay Area Property Acquisition Fund for Equitable Transit-Oriented Development: Feasibility Assessment Report.* Prepared for the Great Communities Collaborative. 2010.

- Denver TOD Fund (operated by Enterprise Community Loan Fund, closed in 2010).

- Puget Sound Affordable TOD Acquisition Loan Fund (currently under development by the Seattle Office of Housing).

- Transit-Oriented Affordable Housing Fund (San Francisco Bay Area), which closed in 2011 with $10 million in top-loss money and an overall capitalization of $50 million.

No known structured funds focus on TOD infrastructure as defined in this study.[151]

KEY FACTORS FOR EVALUATION

Applicability to different types of infrastructure: Because structured funds require repayment, their applicability to TOD infrastructure is limited to activities that have the potential to raise revenue to pay off the fund. Structured funds are typically targeted to land and/or property acquisition and are linked to a specific social mission, like producing affordable housing near transit, where the mission meets the social impact investing criteria for the lenders but still offers a relatively low-risk source of take-out capital. In addition, the typical funding sources for infrastructure, including tax-exempt bonds backed by various funding sources, tend to have lower interest rates than the structured funds and have much lower transaction costs and times.

Approval requirements and legal and political considerations: The use of structured funds does not require voter approval. However, if a public agency is going to contribute top-loss money, it will need political support for the fund. The structured funds in the San Francisco Bay Area and Denver required significant time and effort to build political support for the fund, eventually leading to the participation of a public agency willing to contribute grant money for the top-loss position.

Ease of use: A structured fund is a complex financing tool that would typically be managed by an experienced fund manager according to set rules and procedures. Although a growing body of legal documents can serve as a template for new funds, having an experienced fund manager is often key to the fund's success. Every fund is structured differently depending on its purpose and structure.

USE OF THE TOOL IN PRACTICE

This tool is largely untested for TOD infrastructure. Structured funds are more appropriate for land and/or property acquisition.

Example: Bay Area Transit-Oriented Affordable Housing Fund

Location: San Francisco Bay Area, California

Description: The $50-million Bay Area Transit-Oriented Affordable Housing (TOAH) Fund[152] provides financing to acquire land and develop affordable, transit-oriented housing projects. The fund was formed in collaboration with several public- and private-sector partners, including several community

---

[151] Structured funds are described in greater detail in Appendix C. Fundamentals of Structured Funds.

[152] Bay Area TOAH Fund. "Home page." http://bayareatod.com . Accessed August 31, 2011.

development financing institutions that act as loan originators. The fund has several products, and each partner financing institution offers the same products and terms through the fund.[153] Funding sources included:

- MTC: $10 million.

- Morgan Stanley and Citi Community Capital: $12.5 million.

- Ford Foundation and Living Cities: $3 million each.

- Six community development financial institutions: $8.5 million combined.

- San Francisco Foundation: $500,000.[154]

GETTING STARTED

Implementing a structured fund is a complicated process that could involve multiple public- and private-sector partners and can take years to establish. To apply for money from an existing fund, a project sponsor can usually find specific information on eligibility requirements on the fund's website. As an example of the process to obtain a loan, for the Bay Area TOAH Fund a project sponsor needs to contact one of the originating community development financial institutions, which will determine project eligibility and manage the application and underwriting process for the borrower.[155]

REFERENCES:

Bay Area TOAH Fund. "Home page." http://bayareatod.com . Accessed August 31, 2011.

Torres, Blanca. "Bay Area Agencies Set Up $50 Million Affordable Housing Fund." *San Francisco Business Times.* March 24, 2011. http://www.bizjournals.com/sanfrancisco/news/2011/03/24/mtc-affordable-housing-50-million-fund.html.

---

[153] More information about TOAH is in Chapter III, Section J.

[154] Torres, B. "Bay Area Agencies Set Up $50 Million Affordable Housing Fund." *San Francisco Business Times.* March 24, 2011.

[155] Bay Area TOAH Fund. "Home page." http://bayareatod.com . Accessed August 31, 2011.

G-2.   LAND BANKS

Land banks are public, quasi-public, or private entities that acquire, hold, and manage land to facilitate future development. Originally designed to address issues related to abandoned property, most land banks are public authorities created to acquire vacant or foreclosed properties to stabilize neighborhoods or create affordable housing. A land bank typically acquires tax-foreclosed properties or vacant or underused properties from other government agencies or nonprofits. Properties can also be donated by private owners.

In addition to holding land, land banks can help prepare distressed properties for development by clearing title encumbrances, forgiving property taxes (and thereby removing tax liens), cleaning up environmental contamination, and assembling parcels. For development to actually occur, land banks typically transfer land to private developers with conditions attached that guide how the property will be developed.[156] Properties are usually transferred at below-market value, with preference to nonprofit corporations or entities that will use the property for a public purpose.

Land banks are not a funding or financing tool, but rather an authority with the power to acquire and hold land. Land banks must identify a funding source to pay for acquisitions, maintain the properties, prepare properties for development, and conduct operations and administrative functions. Funding for existing land banks typically comes from the operation of the land bank (e.g., selling other land) and local government contributions.

In principle, a land bank might acquire land for future TOD projects to avoid increases in land value that can result from the introduction of transit service. In this scenario, jurisdictions would create land bank authorities to acquire and hold properties in transit station areas for future affordable or mixed-income housing development or other types of TOD. The challenge with this model is that it requires that the land bank have sufficient resources and funds to acquire and maintain properties over what is often an indeterminate period of time. As an alternative, some groups (usually public or philanthropic) have assembled funds and made them available for land banking by individual property owners. One example is the Twin Cities (Minnesota) Land Acquisition for Affordable New Development Fund, which provides flexible loan financing to assist with acquisition of properties to be held for future affordable housing development.[157] Another example is the North Central Texas Council of Government Sustainable Land Use Development Funding Program, described below.[158]

KEY FACTORS FOR EVALUATION

Applicability to different types of infrastructure: Land banks are targeted to land and/or property acquisition and are often linked to a specific social mission, like neighborhood stabilization or creating affordable housing. While land banks have not been used for TOD infrastructure as defined in this study, the assembly of developable land in station areas could make TOD and associated infrastructure

---

[156] Center for Transit-Oriented Development. "Mixed-Income Transit-Oriented Development Action Guide." http://www.mitod.org/home.php. Accessed September 20, 2012.

[157] Metropolitan Council. "Land Acquisition for Affordable New Development (LAAND)." http://www.metrocouncil.org/services/LAAND.htm. Accessed August 24, 2011.

[158] More details about the potential for land banking for TOD can be found in Chapter IV, Section C.

projects more feasible. Land banking could be useful for infrastructure projects when land assembly is needed as part of a larger TOD or station area project.

Approval requirements and legal and political considerations: Establishing a land bank would require local enabling legislation. State enabling legislation might also be necessary depending on the location.

Application for strong and weak real estate markets: This tool is typically used to acquire land in weaker real estate markets.

Capacity and scale: A land bank's capacity depends on the availability of funds for acquiring and maintaining property.

Ease of use: Establishing and maintaining a land bank involves significant administrative and operational tasks. Land banks can be established either as public agencies that share staff with other city departments or as separate entities with their own board of directors.

OTHER LIMITATIONS OF THE TOOL

Acquiring and holding properties involves significant risk. Because land banks often hold property while waiting for an appropriate development opportunity, they can incur significant carrying costs, including for maintaining the property.

USE OF THE TOOL IN PRACTICE

Example: North Central Texas Council of Government (NCTCOG) Sustainable Land Use Development Funding Program: Land Banking Program

Location: North Central Texas

Description: NCTCOG's land banking program was established to help local governments assemble parcels for redevelopment in the future.[159] In 2006, the program funded four land banking projects, including one at Addison Circle, a plaza at a Dallas Area Rapid Transit station. The NCTCOG land banking program will not condemn properties or pay for relocation costs or the costs associated with a land purchase.[160]

GETTING STARTED

New land banks are typically established by a local government agency, such as a city or a municipal planning organization, although some land banks have been established by nonprofit organizations. The process for establishing a land bank includes identifying a funding source to acquire, hold, and manage land.

---

[159] NCTCOG. "Sustainable Development Funding Program: Landbanking Projects." http://www.nctcog.org/trans/sustdev/landuse/funding/lbank.asp. Accessed August 31, 2011.

[160] More details about this program are in Chapter IV, Section C.

REFERENCES

Alexander, Frank. *Land Bank Authorities: A Guide for the Creation and Operation of Local Land Banks.* Local Initiatives Support Corporation. 2005. http://www.lisc.org/content/publications/detail/793.

Center for Transit-Oriented Development. "Mixed-Income Transit-Oriented Development Action Guide." http://www.mitod.org/home.php. Accessed September 20, 2012.

Metropolitan Council. "Land Acquisition for Affordable New Development (LAAND)." http://www.metrocouncil.org/services/LAAND.htm. Accessed August 24, 2011.

NCTCOG. "Sustainable Development Funding Program: Landbanking Projects." http://www.nctcog.org/trans/sustdev/landuse/funding/lbank.asp. Accessed August 31, 2011.

G-3.   REDFIELDS TO GREENFIELDS

Redfields to greenfields[161] is not a funding source or financing tool but a concept, similar to a land bank, for converting underused or distressed properties into an asset. The concept involves a local government agency acquiring underused properties in an area and converting them into parkland. Privately held properties are the focus of the concept, but publicly held properties could also play a role. Converting underused or distressed properties into parks can increase the property value of adjacent parcels, thereby creating a value capture opportunity to fund additional acquisitions and conversions.[162] The redfields to greenfields method of converting properties is an emerging tool at this point, although there are some places where programs have begun or are beginning.

Redfields to greenfields is not tied to a particular funding source; in fact, implementing the concept would require a significant funding source to be identified. Proposals for this concept have suggested that low-cost loans could be made from a new land bank and parkland acquisition fund established by the nation's banking system and led by the Federal Reserve, U.S. Treasury, and Federal Deposit Insurance Corporation. The proposals also suggest using other financing tools such as tax credits leveraged with local equity capital, although these sources would likely need to be repaid as well.[163]

Additional funds would be needed to pay for the improvements to convert the properties into parklands. Redfields to greenfields could rely on some of the other funding sources discussed in this report. In some instances, if funding for the conversion to parkland is not immediately available, the acquired properties could be held in a land bank for conversion to parkland in the future.

KEY FACTORS FOR EVALUATION

Applicability to different types of infrastructure: This concept is focused on parks, which could include parks that are part of a mixed-use TOD.

Approval requirements and legal and political considerations: Municipalities will need to determine the source of capital, the entity that will manage the program, and how it will operate.

Application for strong and weak real estate markets: This tool is focused on underused or distressed properties that are most likely to be present in weak real estate markets, but it could also be applied to weaker submarkets in a strong metropolitan real estate market.

Capacity and scale: Because this concept is not a funding or financing tool, this factor is not applicable.

Ease of use: The redfields to greenfields concept requires a complex series of steps to implement, likely including the acquisition of foreclosed or other distressed properties.

---

[161] "Redfields" refers to underused properties, and "greenfields" refers to parks and open space.

[162] See Sections E-1 through E-4 in this appendix for more information on value capture tools.

[163] Redfields to Greenfields. "Home page." http://rftgf.org/joomla/. Accessed August 29, 2011.

OTHER LIMITATIONS OF THE TOOL

Availability of funds to pay for the acquisition of properties and improvements would limit the use of this tool. In addition, because the tool relies on acquiring underused properties, it would be subject to the land acquisition and assembly challenges typical of transit station areas, where appropriate parcels can be relatively rare.

USE OF THE TOOL IN PRACTICE

This tool has not been used extensively; it is primarily in the concept stage.

Example: Atlanta BeltLine

Location: Atlanta, Georgia

Description: The Atlanta BeltLine project[164] is a $2.8-billion redevelopment project that will create a network of public parks and multiuse trails along a 22-mile transit corridor. The parks and greenbelt component of the BeltLine Redevelopment Plan calls for the acquisition of 585 to 625 acres of redfields and other open space sites for $480 to $570 million, and the development of 260 to 300 acres of parks and trails for $275 to $340 million over 25 years. The project is in part intended to spur TOD in the approximately 3,000 acres of underused or idle industrial land around the BeltLine. Proposed funding sources include BeltLine Tax Allocation District funds, a capital campaign, Park Opportunity Bonds, Department of Watershed Management funding, and federal funding for trails.

GETTING STARTED

A nonprofit organization called Redfields to Greenfields has conducted a series of studies focused on Atlanta, Cleveland, Denver, Miami-Dade, Philadelphia, and Wilmington to analyze the potential effects of the program on these cities and lay out the process from acquisition to eventual sale of the land.[165]

REFERENCES

Atlanta BeltLine. "Home page." http://beltline.org. Accessed August 29, 2011.

Refields to Greenfields. "City Studies." http://rftgf.org/joomla/index.php?option=com_content&view=article&id=149&Itemid=50. Accessed July 30, 2012.

Redfields to Greenfields. "Home page." http://rftgf.org/joomla/. Accessed August 29, 2011.

---

[164] Atlanta BeltLine. "Home page." http://beltline.org. Accessed August 29, 2011.

[165] Refields to Greenfields. "City Studies." http://rftgf.org/joomla/index.php?option=com_content&view=article&id=149&Itemid=50. Accessed July 30, 2012.

G-4. NATIONAL INFRASTRUCTURE BANK

Discussion about creating a national infrastructure bank (NIB) dates to the mid-1990s, when a backlog of infrastructure projects and the lack of appetite for selling public debt to fund construction costs were perceived as bottlenecks to the United States' long-term growth. Since the 2008 financial crisis, the fragile fiscal position of several states and municipalities has revived the discussion.

An infrastructure bank provides credit assistance for infrastructure projects through loans, guarantees, or other credit assistance. Infrastructure banks can be independent federal agencies, state government entities, or private-sector or nonprofit corporations. The key difference between an infrastructure bank and a commercial bank or private-sector infrastructure fund is that infrastructure banks are government-established.[166] Examples include the European Investment Bank and, in the United States, state infrastructure banks.

Four infrastructure bank bills were proposed in the 112th Congress: S. 652[167], S. 936[168], H.R. 402[169], and H.R. 3259[170]. S. 936 would create a fund in DOT, while the others would create a wholly owned federal government corporation. Though an NIB does not currently exist, if established, it could help states or other government entities access financing for infrastructure projects.

KEY FACTORS FOR EVALUATION

Applicability to different types of infrastructure: All the bills propose offering financing to transportation projects, among other sectors.[171] Without an established bank, the type of transportation project components that could have access to financing is unknown. Bill S. 652 offers more specificity on the size of eligible projects. According to this, the estimated cost of individual projects would have to be at least $100 million or, for rural infrastructure projects, $25 million. Bill S. 936 has a broader definition of eligible projects and defines them as "activities included in a regional, State, or national plan" and "transportation related."

According to the White House, loans made by the bank would be matched by private-sector investments or money from local governments so that the infrastructure bank would provide no more than half of the total funding. Each project would identify funding sources to help ensure repayment of the loan.

---

[166] Copeland, C., Mallett, W., and Maguire, S. *Legislative Options for Financing Water Infrastructure.* Congressional Research Service. 2012.

[167] On March 17, 2011 the bill was referred to the Senate Committee on Finance, but no further action was taken as of July 31, 2012.

[168] On May 10, 2011 the bill was referred to the Senate Committee on Commerce, Science, and Transportation, but no further action was taken as of July 31, 2012.

[169] On March 23, 2012 the bill was referred to the Subcommittee on Domestic Monetary Policy and Technology, but no further action was taken as of July 31, 2012.

[170] On January 12, 2012 the bill was referred to House Subcommittee on Insurance, Housing and Community Opportunity, but no further action was taken as of July 31, 2012.

[171] S. 652, H.R. 402, and H.R. 3259 include other sectors such as water and energy.

Sources for repayment could include tolls, user fees, or other dedicated state or local government sources.[172]

Capacity and scale: Based on the information available, it is likely that the minimum project size would be $25 million to $100 million, which might exceed the project size in small communities. One alternative for small communities to reach the minimum project size would be to propose regional projects, which would require coordination across multiple agencies that can be time consuming but could increase their opportunities to access financing.

Timing and lifecycle: According to S. 652, the interest rate on the loans could not be less than the yield on U.S. Treasury securities of similar maturity, and the term of the loans could not exceed 35 years. According to S. 936, the term of the loans could not exceed 90 percent of the estimated useful economic life of the asset being financed. According to H.R. 3259 the term of the loans could not exceed 35 years or 90 percent of the useful life of the asset, whichever is less.

Ease of use: All bills propose that the projects be subject to criteria that include the benefits generated by the project, its funding gap, and committed sources.

USE OF THE TOOL IN PRACTICE

Although a national infrastructure bank does not exist, an example of a state infrastructure bank appears in Section B-7 of this appendix.

GETTING STARTED

N/A

REFERENCES

Compton, Matt. "Five Facts About a National Infrastructure Bank." The White House Blog. November 2, 2011. http://www.whitehouse.gov/blog/2011/11/03/five-facts-about-national-infrastructure-bank.

Copeland, Claudia, Mallett, William, and Maguire, Steven. *Legislative Options for Financing Water Infrastructure.* Congressional Research Service. 2012. http://www.fas.org/sgp/crs/misc/R42467.pdf.

Mallett, William J., Steven Maguire, and Kevin R. Kosar. "National Infrastructure Bank: Overview and Current Legislation." *Congressional Research Service.* December 14, 2011.

Likosky, Michael. "Banking on the Future." *New York Times,* July 13, 2011. http://www.nytimes.com/2011/07/13/opinion/13likosky.html.

---

[172] Compton, M. "Five Facts About a National Infrastructure Bank." The White House Blog. November 2, 2011. http://www.whitehouse.gov/blog/2011/11/03/five-facts-about-national-infrastructure-bank.

Planners and community developers have pursued structured funds as a property acquisition tool to support affordable housing development, often with a focus on transit-accessible locations.[1] This appendix includes an overview of structured funds used in this way, including:

- An introduction to the structured fund concept, including why these funds are well suited to fund property acquisition for affordable, transit-oriented housing.

- How mission-driven funds get organized and are managed.

- Steps for getting organized to start a structured fund.

This appendix expands on the discussion in Section G-1 of Appendix B to provide more in-depth information about this tool. However, because structured funds can take various forms, even this more detailed discussion provides only a brief introduction to this complex topic. Nevertheless, it should help readers understand where, when, and how a structured fund can be used to finance a variety of purposes, as well as who should be involved in forming a fund and what kind of process is necessary.

A. INTRODUCTION TO STRUCTURED FUNDS

The term "structured fund" refers to a kind of loan fund that pools money from investors with different levels of risk tolerance and different demands for investment returns. By blending capital, which typically comes from various sources including the public, philanthropic, and private sectors, fund managers can use money from investors willing to take higher risks and lower returns to leverage investments from investors who require higher returns but want less risk. Community development financial institutions and other community development intermediaries have used structured funds to access larger pools of capital than might otherwise be possible while simultaneously lending money at below-market interest rates.

Structured funds have a dedicated purpose, which is clearly defined prior to fund formation, and they are managed by professionals with fund formation and loan underwriting experience. Each fund is its own legal entity and is organized, or "structured," around a credit agreement that defines the level of risk each investor will take on and how money will be dispersed to investors as loans are repaid. The pooled funds are organized into a capital stack with layers that represent differing levels of risk and return expectations. Multiple investors can be situated in each layer in the stack. Investors willing to take the greatest risk for the lowest expected return are at the top of the stack and will be the first to absorb any losses. Typically these investors are mission-driven public or philanthropic entities. Conversely, investors who want to take the least risk but achieve the highest returns, such as banks, are at the bottom of the stack, where their money is essentially protected by the higher-risk investors. Credit agreements also structure the "payment waterfalls" that stipulate who in the stack will be paid what amounts and in what order. Unlike other kinds of investment vehicles where high risk equates to

---

[1] Using structured funds as a property acquisition tool has led some people to use the term "acquisition fund" interchangeably with "structured fund." While structured funds can be a type of acquisition fund, there are many other types of acquisition funds, and structured funds can be used for purposes other than property acquisition.

high return, in these community development-driven structured funds, the investors who take the least risk get paid first while the highest-risk investors get paid last and sometimes expect no or minimal interest on their investment.

Structured funds are "closed ended," meaning they have a set term establishing when investors will be repaid, which helps mitigate their risk. Structured funds also tend to focus on a narrow band of loan types. This focused lending also helps mitigate risk because investors are able to better understand the assets to which the fund will make loans.

Structured funds have four main benefits:

- They can create a relatively large fund by leveraging multiple investors.

- They can provide subsidized interest rates by blending different return expectations from different investors.

- Loans can be underwritten relatively quickly and as needed, similar to a bank loan and unlike public-sector or philanthropic grants that are generally disbursed on fixed cycles.

- Clear coordination among investors allows borrowers to go to one source to obtain financing, rather than having to piece together funding from multiple sources. [2]

Structured funds also present five major challenges:

- These funds rely heavily on the first, or "top loss," investor who is willing to absorb the greatest risk and take the lowest (or no) return. This source of credit assistance typically, although not necessarily, comes from a public-sector investor who is willing to put its money in the fund as a grant or a no-interest loan. The size of public-sector top loss will affect the size of the fund or the risk profile of the loans. A fund can be structured without public money, but the fund might be smaller or take on a narrower range of project types. Without a significant contribution by a top-loss investor, the fund will not work.

- Because the credit agreement at the core of each fund is always complex and involves considerable negotiations with all parties involved, structured funds have significant start-up costs. These transaction costs will influence whether a structured fund is appropriate.

- Although structured funds are set up to accommodate multiple investors, too many investors can be problematic. Trying to negotiate the fund structure among too many parties with differing goals and priorities can make for an overly complex credit agreement and create inefficiencies in fund operations.

- The funds need to meet a specific demand. Structured funds are a particular kind of loan tool, and they might not be suited for some potential borrowers' needs, depending on the market and the availability of other money.

---

[2] Lal Schmidt, Deidre. *Strategic Acquisition Fund for Transit Oriented Development (TOD): Understanding the National Experience, Exploring the Needs and Opportunities in the Twin Cities Region.* The Family Housing Fund. December 2011.

- The funds need to be designed with some flexibility. However, a fund that allows for some refinement in underwriting criteria and loan products over time while also providing certainty for investors is difficult to structure. [3]

B. USING STRUCTURED FUNDS FOR PROPERTY ACQUISITION

Property-acquisition structured funds differ from other kinds of property-acquisition funds in several ways. First, because structured funds have a specific end date, the loans are relatively short term (typically five to 10 years) and are not appropriate for buying and holding property for indefinite periods. Second, capitalization for the structured funds comes from investors who have clearly defined risk and return expectations. Most other acquisition funds with a public purpose or social mission are tied to public entities or nonprofits such as land trusts or land banks, where the capital sources come from donors, public funds, or other sources expecting only a minimal return, if any.

Just as a structured fund is not the only kind of property-acquisition fund, property acquisition is not necessarily the only function for a structured fund. However, mission-driven structured funds lend themselves to property acquisition particularly well if the funding is being used to "bridge" longer term financing and/or if the loan can be paid back within the fund's term. In addition, real estate is a well-understood asset. As security for the loan from the structured fund, real estate can help the investors, fund manager, loan underwriters, and borrowers to clearly understand the potential risks and rewards associated with the loans being made. The most successful mission-driven structured funds that have been established to date have focused only on real estate-related lending.

In theory, a structured fund could be established to underwrite small-business loans or other activities, such as infrastructure projects, that might be related to TOD implementation. However, because the risk profiles of these other activities are different from real estate lending, different investment sources and underwriting criteria would be required. Because structured funds operate best with a relatively narrow and clearly defined risk profile, a single fund is unlikely to make loans to a wide range of activities with widely varied risk profiles. Communities seeking tools to support activities other than property acquisition should either consider creating a separate structured fund for those activities or look to other kinds of funds.

C. STRUCTURED FUNDS AND TOD

A 2006 report showed that transportation costs are highly sensitive to location and that, on average, households living in walkable, transit-rich locations paid about half the transportation costs of households living in suburban, automobile-oriented communities. [4] Because transportation costs are the second-highest household expenditure after housing, lower-income households' access to transit could have a profound impact on their personal finances. The less these households have to spend on transportation, the more money they have to pay for other expenses such as healthy food, medical care, and education. A 2007 report estimated that by 2030, 15 million households would want homes near

---

[3] Ibid.

[4] Lipman, B. J. *A Heavy Load: the Burden for Housing and Transportation Burdens Working Families*. Center for Housing Policy. 2006. http://www.cnt.org/repository/heavy_load_10_06.pdf.

transit, with about 40 percent of this demand coming from low- and moderate-income households. [5] The findings from both of these reports made a clear case that housing and transportation policy are inextricably linked. If both support the production of affordable housing at transit locations, then transit can more effectively serve the needs of households at all income levels.

Although many tools are designed to preserve and build mixed-income housing near transit, during the housing boom of the early 2000s, many organizations began focusing on land and property acquisition as a key strategy to ensure that TOD included homes for low- and moderate-income households. This issue was most pronounced in regions with strong real estate markets where developers and land speculators were targeting prime development opportunity sites near transit, including Denver and the San Francisco Bay Area.

Structured funds emerged as a promising acquisition financing strategy based on the experience of the pioneering New York City Acquisition Fund, designed to assist local affordable housing developers with funds for land acquisition in a very competitive market. New York City and Enterprise Community Partners, along with several foundations and banks, developed a structured fund that provided bridge loans for affordable housing developers who needed to purchase property before having full project financing. The New York City Acquisition Fund leveraged $265 million through a complex structure including a mix of public, philanthropic, and private resources. Since 2006, when the New York fund closed, many other mission-driven acquisition funds have been created, but only two—the Denver TOD Fund, serving the city of Denver, and the Transit Oriented Affordable Housing (TOAH) Fund, serving the San Francisco Bay Area—were created specifically to support affordable housing and community facilities near transit. The other mission-driven acquisition funds are also primarily focused on affordable housing production but not necessarily in locations near transit stops.

D. ESTABLISHING THE NEED FOR A STRUCTURED FUND

One of the most important preconditions for creating a structured fund is having a clear purpose or need that the fund will address. Both the potential investors and borrowers must agree that this need exists, and the fund's governance, loan products, underwriting criteria, and risk and return expectations will be organized around this purpose or need. Establishing this need requires commitment over a long time horizon. For example, the Bay Area TOAH Fund required a two-year, four-stage process extending from fund inception to fund closing. A significant portion of the process was devoted to establishing a clear need for a real estate-based acquisition structured fund. Although the four-stage process is described below as discrete activities, in fact, there was considerable overlap both in terms of timing and purpose among the four.

- The first and most critical stage in the TOAH creation process was to build a multisector relationship among all of the key groups necessary to make the fund successful. These actors included major regional advocacy groups, community foundations, the affordable housing community, and the regional agencies whose programs the fund would support and who would ultimately provide the top loss money for the fund. One of the most important actors was the Great Communities

---

[5] Center for Transit-Oriented Development. *Realizing the Potential: Expanding Housing Opportunities Near Transit.* 2007. http://www.reconnectingamerica.org/resource-center/books-and-reports/2007/realizing-the-potential-expanding-housing-opportunities-near-transit-2.

Collaborative (GCC), whose core membership included the region's largest environmental, social justice, and affordable housing nonprofit organizations; a national TOD intermediary; and two community foundations. Over the two-year fund formation period, the GCC was the primary convener and facilitator of the process, bringing all stakeholders along at each critical juncture. Without the GCC in this role, the fund would not have been created.

- During the second stage of the process, the GCC engaged the Center for Transit-Oriented Development to prepare a report that established a clear need for a property acquisition structured fund, articulated the fund's overarching mission and potential governance structure, and established a preliminary target geographic area for loans. This report became, in essence, the fund prospectus. Being able to clearly articulate how the fund's mission met the region's need was critical to attracting investors. The report reflected multiple discussions with key stakeholders, including the advocacy groups and regional agencies whose goals and programs would be broadly supported by the fund (the mission); the regional agency that eventually contributed the top-loss money; the community foundations that invested critical second-loss capital (the investors); and end users (the borrowers). When this report was released, it represented a clear consensus across all parties.

- The third phase in the process was to identify a source of top-loss funding. Because the fund was regional, the source had to have a regional mission. Other funds, including Denver's, were established based on top-loss support from an individual city and thus required that all funds be spent within that jurisdiction. Because regional funding sources are limited, the only likely candidate was the MPO, MTC. In the final analysis, the fund's need was tightly framed to align with MTC's long-term objectives. Although all MPOs are primarily focused on providing transportation investments to support future regional growth, MTC realized that it would never have sufficient funds to support future growth if that growth continued to stretch outward. Instead, MTC determined that investing in tools to help refocus growth into the region's core would lead to a more cost-effective strategy for fulfilling the region's future transportation needs. MTC also recognized that by supporting affordable housing construction near transit, it was supporting more consistent transit ridership and ultimately reducing demand for additional highway capacity, given that affordable-housing residents have relatively low rates of automobile ownership and high rates of transit ridership. As part of MTC's agreement to contribute top-loss money, the organization's board stipulated a timeframe in which the fund needed to be capitalized (progress needed to be made within 14 to 18 months) and an expected leverage ratio (i.e., for every $1.00 MTC invested in the fund, it wanted the fund manager to raise at least $2.50 to $3.00 in additional capital).

- Once MTC had agreed to provide the top-loss investment, the fund moved into its fourth and final stage, implementation. At this point, the GCC and MTC worked together to hire a fund manager. Once the manager was on board, it became the manager's responsibility to write a detailed business plan and, ultimately, the credit agreement that created the fund's actual structure. Only at this point in the process was it possible to determine, based on size of the initial top-loss grant and the nature of the specific investors, the overall size of the fund, interest rates, specific loan products, detailed underwriting criteria, and other elements.

- A similar needs assessment process has been taking place in the Minneapolis-St. Paul (Twin Cities) region in Minnesota. Although no entity in the Twin Cities is exactly analogous to the GCC, the region still has considerable cohesion among the many actors promoting equitable TOD. Several organizations partnered to retain a consultant to help the region explore the possibility of creating a structured fund to support property acquisition along transit corridors and create more housing choices and better access to community services.

In a report prepared for the Twin Cities, the initial questions were framed around three issues:

- Whether it should be the region's highest priority to invest affordable housing funds near transit versus targeting investments to other kinds of locations.

- Whether property acquisition is the most significant development challenge for producing affordable housing near transit or whether funding other development activities (e.g., construction or pre-development planning) should have higher priority.

- Whether a structured fund was the right tool for meeting the region's needs.[6]

This report did not identify a clear need for a structured fund. The report authors identified multiple existing funding sources for property acquisitions in the Twin Cities. While the report did find a clear need for more permanent financing sources for affordable housing and did not rule out the possibility of creating a fund, the recommendations called for an ongoing exploration of other mechanisms to support gaps in funding for affordable housing while continuing to address ways in which a structured fund could be organized to meet the region's needs as they become more clearly defined.

E. TOP-LOSS MONEY AND SIZING STRUCTURED FUNDS

Top-loss money is critical to a mission-driven structured fund because it provides the basic credit assistance, or risk reduction, necessary to protect and therefore attract a wider range of investors. One key element of the credit agreement structuring any fund is how losses will be distributed among the various investors. Typically, the top-loss investor takes the greatest percentage of any losses, and the next tier or tranche takes a different but also larger share of loss than the senior investors (see Exhibit C-1). The capital stack works most effectively when there is sufficient capital in the top-loss and mid-level tranches to protect the senior investors, who sit at the bottom of the stack. Typically, senior investors will want enough tiers and capital between their investment and the top-loss investor to ensure that the senior debt has a maximum exposure of 55 to 60 percent of the total fund loan to value ratio.[7]

---

[6] Lal Schmidt, Deidre. *Strategic Acquisition Fund for Transit Oriented Development (TOD): Understanding the National Experience, Exploring the Needs and Opportunities in the Twin Cities Region*. The Family Housing Fund. 2011.

[7] Personal communication with Brian Prater, Managing Director for the Western Region, Low Income Investment Fund, by Dena Belzer, Strategic Economics, on February 9, 2012.

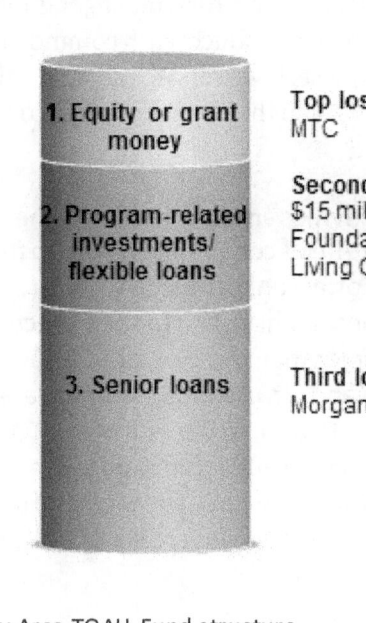

Top loss (public sector): $10 million from MTC

Second loss (foundations and CDFIs): $15 million from six CDFIs and Ford Foundation, San Francisco Foundation, and Living Cities

Third loss (banks): $25 million from Morgan Stanley and Citi Community Capital

Exhibit C-1. Bay Area TOAH Fund structure.
Source: Bay Area TOAH. "Bay Area's Transit-Oriented Affordable Housing Fund." Presented at Affordable Housing Week. May 2011.

This relationship between the top-loss and other investors in the fund establishes the ultimate size of the fund and is the reason that a fund's size cannot be established before identifying the source and magnitude of the top-loss tranche. Structured funds are also often characterized in terms of "leverage"—i.e., the amount of investment that came into the fund relative to the initial top-loss contribution. While it is possible to set a goal for leverage before establishing the fund, the actual leverage ratio depends on the detailed credit agreements among all the investors and cannot be calculated until after the fund is closed and the total capital is known. For example, MTC exceeded its goal to raise at least $2.50 to $3.00 in additional capital for every $1.00 MTC invested in the fund. In fact, the fund manager was able to raise $4.00 for every $1.00 MTC invested.

Because the top-loss position is at such high risk, the mission-driven funds typically rely on grant funding from a public agency to fill this tier in the capital stack. From MTC's perspective, contributing $10 million in top-loss funds to the TOAH Fund had a significant and tangible benefit because the grant money was being leveraged in a way that both amplified the impact of the initial $10 million and had the potential to produce many more affordable housing units than MTC could generate through direct investment. Given this cost-benefit equation, MTC determined that the risk of contributing $10-million that might not get repaid was far outweighed by the potential benefits.

F. FUND ORGANIZATION AND MANAGEMENT

FUND ACTIVITIES AND LOAN PRODUCTS

Establishing the overarching need or mission of a structured fund and defining the specific activities and loan products the fund will undertake are not necessarily the same thing. Decisions about fund activities, loan products, and targeted borrowers are closely linked to the fund's business plan and credit structure. Therefore, these detailed decisions are typically made during the business-planning phase of fund formation once the top-loss money has been identified and a fund manager is raising capital. However, having a relatively narrow set of activities and loan products with common underlying risk factors established at an early stage is important. Otherwise, investors might not be able to agree on such issues as fund governance and underwriting criteria, and they might not feel confident in their expected returns—whether financial or mission-related.

Different fund structures can yield different kinds of activities and loan products; there is no one-size-fits-all approach. The Denver TOD Fund's loan activity reflects conditions specific to the Denver region, investor requirements, and borrower needs. This fund offers only one kind of loan (a revolving line of credit), lends to only one borrower (the Urban Land Conservancy), and funds projects only in the city of Denver. On the other hand, the TOAH Fund, which is larger than the Denver TOD Fund, offers five different loan products, is open to multiple borrowers, and can use funds in designated growth areas throughout the nine-county San Francisco Bay Area. In addition, 85 percent of the TOAH Fund is for housing projects, and 15 percent is for community facilities including daycare centers, health clinics, and neighborhood food stores. However, both funds are focused on real estate-related lending, even for non-residential-related borrowers such as childcare centers. All loans are collateralized by the underlying real estate asset.

FUND GOVERNANCE AND MANAGEMENT

Mission-driven structured funds are similar in many ways to community development financial institutions. Both entities are set up to address a broad social mission, but both also require specialized skills to ensure that the financial services being offered support the broader mission. This need to address a social mission while having a sound business model has resulted in communities establishing a two-tiered approach to managing their funds. The first tier of governance is usually an oversight or advisory committee composed of individuals who might not be experts in fund management but who represent the fund's broad social interests. The members of this advisory group can be defined in the fund's needs assessment or prospectus document. The advisory committee generally meets periodically to review loans to determine how consistent the loans are with the fund's mission, and to give the fund manager and loan underwriter direction on future activity. This group acts as a watchdog to be sure the fund is fulfilling its intended mission, but it is not involved in the fund's actual lending activities. Separating the advisory function from the loan function ensures that loan decisions are made on the merits of each borrower and that political or other non-financial factors do not play a role in making the loan decisions. Without this clear separation between mission and operation, funds might be more likely to make bad loans, putting investor funds at greater risk, and thus making the fund less attractive as an investment vehicle.

Given the complexity of structure funds, it is essential that they be managed by a professional fund manager or management team with experience in structuring credit agreements and underwriting loans for the fund's specified types of activities. While the fund manager has the main responsibility for the fund's structuring and capitalization, he or she is generally hired only after a clear need for a structured fund has been established and a top-loss investor has been identified. Once hired, the fund manager is then responsible for developing a detailed business plan for the fund, including identifying the fund's specific activities, loan products, eligible borrowers, underwriting criteria, and loan originators. Having a clear, well-organized business plan facilitates developing the credit agreements that become the legal structure for all of the management responsibilities, risk allocation, and repayment order. Various structured funds have taken different approaches to identifying who should sit on a credit committee that decides which loans should be considered. In most cases, the credit committee includes finance experts and fund investors but not policy-makers, such as elected officials. This arrangement helps ensure that loan decisions are made on the deal's individual merits and avoids conflicts of interest between protecting the fund's investments and other political priorities.

FUND INVESTORS AND CAPITAL SOURCES

The success of any structured fund depends on the fund manager's ability to attract capital from diverse sources that encompass a spectrum of risk/return profiles. This pooling strategy is central to making mission-driven structured funds work for their intended borrowers because the blending of risk and the payment waterfalls encourage more risk-averse investors to put capital into a fund that will both lend at below-market interest rates and make loans that appear riskier than conventional commercial real estate loans. However, to work, the structure must include mission-driven investors willing to take greater risk and earn lower expected returns.

As explained above, the top-loss layer in the capital stack typically comes from a government source whose tolerance for risk is higher because it is primarily concerned with achieving the public purpose that the fund will address, such as producing affordable housing. Some funds have been started with a loan guarantee rather than actual capital. In this case, a governmental or philanthropic entity assumes the liability for a certain percentage of potential losses from the fund but does not invest actual capital. While this approach can provide the credit assistance necessary to attract other investors, it also decreases the amount of money that is available for lending. A loan guarantee cannot be lent out, but in theory, some portion of the top-loss capital can, although only under conditions acceptable to the more senior investors.

The next tier or tiers in the fund represent investors who have a higher tolerance for risk than conventional investors. Like government, these investors want to address a social mission, but they are also expecting some return. Typical investors for the second-loss position include foundations making program-related investments. Under the Internal Revenue Code, foundations are allowed to invest a certain portion of their charitable giving as program-related investments where the investment supports the foundation's tax-exempt purpose and where the primary objective of the investment is not to produce significant income or benefit a political purpose. [8]

Community development financial institutions have also been important investors in structured funds for much the same reason as foundations. The funds support their overall mission and also allow some return on investment. In some cases, a community development financial institution has both been an investor in the fund and acted as the fund's manager.

Senior debt sits at the bottom of the stack and typically comes from commercial sources such as banks and insurance companies. Banks are sometimes motivated to invest in structured funds to meet their obligation under the Community Reinvestment Act to help meet the credit needs of the communities in which they operate.

---

[8] IRS. "Program-Related Investements." http://www.irs.gov/charities/foundations/article/0,,id=137793,00.html. Accessed September 2011.

Capital stacks can be organized in several different ways relative to loan underwriting activities. One approach is to bring investors in to evaluate each deal and determine whether they want to participate in a loan. While this approach offers investors considerable flexibility, it is also inefficient and time-consuming, does not maximize the opportunity for leverage within the capital stack, and can result in fewer loans. By contrast, the TOAH Fund is truly a blended fund where all of the investors put their money into a single pool. Each loan is made by drawing down on the entire pool. Although investors might not have the same flexibility with each loan, there is less risk overall, and because all investors are represented on the TOAH Fund's loan committee, they still have considerable control over what loans are being made.

The fund's size is ultimately determined by the amount of top loss, and thus credit assistance, the fund can offer. The size of this investment must be big enough to warrant the cost and complexity of creating a structured fund. Although there is no rule of thumb about a minimum structured fund size, discussions about the amount of top loss are a critical part of the first phase of fund formation when the fund's sponsors or advocates are evaluating creating a structured fund versus another type of loan fund to meet their community's needs.

One of the central challenges to creating a structured fund is identifying a source of top-loss money. The source of top-loss money will be a key determinant of the size of the fund and other attributes. For example, if a city is the source of top-loss funding, then loans can be made only within that jurisdiction; there would have to be sufficient demand for loans in that city to warrant creating a fund. Where a fund is intended to support a transit corridor that traverses multiple jurisdictions, finding a top-loss investor is more challenging. In theory, multiple jurisdictions could pool money from a variety of sources, including tax increment revenue. However, the funds would have to be fully assembled (i.e., the top-loss money could not be capitalized through commitment of future revenue such as future tax increment). The top-loss money also could not come from the proceeds of bond sales. Senior investors are likely to view the top-loss money protected by future tax increment or bond sales as too insecure to protect their investments. Another option, as in the TOAH Fund, is for an MPO or other regional organization, rather than a city, to hold the top-loss position.

GETTING STARTED—STEPS FOR ESTABLISHING A FUND

WHO SHOULD BE INVOLVED IN STARTING A STRUCTURED FUND?

The players involved in starting a structured fund depend largely on whether the source of top-loss capital has been identified. In communities where no source of top-loss funding has been identified, a group of supporters needs to coalesce around a process for establishing the fund's need, identifying the potential source(s) of top-loss money, and making the case to prospective investors that the fund is both warranted and a good investment. The entity that leads this effort will vary from community to community. However, whatever group coalesces around creating the fund should also reflect the fund's potential geographic reach. For example, if the fund will serve a region, it should have representation from a variety of regional groups or entities, while a fund serving only one community could be composed entirely of people from that community. In the Bay Area and the Twin Cities, local foundations played a key role at this point in the process by providing financial support, convening key stakeholders who could help shape the fund's mission, and reaching out to other investors, including national foundations. In Denver, the process began a little differently as the Enterprise Community Partners, a national community development intermediary as well as a foundation, took the lead. When

Enterprise was considering creating a fund, there were no local foundations specifically interested in affordable TOD.

Having a strong local or regional capability to bring stakeholders together, prepare a compelling case for a structured fund, and identify an appropriate source of top-loss money are all important to providing investors, including national foundations and community development financial institutions as well as senior investors, with the confidence that the community can create a viable and well-structured fund. Without national sources of capital, the existing funds would likely not have been able to raise sufficient funds to meet their leverage goals.

In the Bay Area, the GCC, in partnership with the San Francisco Foundation and the Silicon Valley Community Foundation, was the pivotal entity that brought together all of the right actors to establish both the need and the mission for the TOAH fund. The GCC hired consultants to help with critical aspects of this process. The Center for Transit-Oriented Development prepared the fund's needs statement and prospectus, and Imprint Capital Advisors helped select the fund manager.

In some cases, the public entity willing to invest top-loss money initiates the structured fund. For example, Los Angeles County, California, developed a structured fund to assemble land for affordable housing projects. Having the top-loss investor take the lead in forming a fund can greatly simplify the process but does not negate the need to establish a clear, tightly focused mission. Investors still need confidence that their investment will be secure within their risk/return parameters and understanding of the types of activities the fund will undertake.

HOW TO ESTABLISH THE NEED

Each community will need to do a detailed assessment of the local barriers to developing TOD and make an initial determination as to whether a structured fund is an appropriate vehicle for addressing one or more of the most pressing barriers. This process will need to include potential top-loss investors, local community foundations, likely fund borrowers, and possible fund managers. Tying the fund's identified need to the objectives of both the investors and the funders is critical to establishing a structured fund. Although structured funds do not necessarily have to invest in property acquisition, it is the most common use of the mission-driven structured fund tool. If a community wants to use this tool for another purpose, there will be a considerable learning curve, especially for the investors. Outreach, research, and education will be critical to ensuring that investors are comfortable with the risk/return profiles available from the fund's proposed assignment of risk and payment waterfall.

HIRING A FUND MANAGER

Fund managers are typically community development financial institutions or private-sector financial advisors with experience in structuring pooled loan funds and/or forming real estate investment funds. Having the right level of experience and expertise is critical for attracting investors and ensuring that the fund performs its mission while providing the expected return for its investors. However, it is important not to hire the fund manager too early in the process. Those organizations that are best suited to manage a fund might not have the skills or even the local knowledge to identify the fund's need and mission and the top-loss investor(s). This effort is best performed by local organizations whose objectives match the fund's purpose and who can rally the necessary political and social capital to attract a top-loss investor and build credibility with other non-local investors.

LESSONS LEARNED FROM THE BAY AREA TOAH AND DENVER TOD FUNDS

Starting a mission-driven structured fund is a difficult task that requires partnerships, focus, and a clear set of fund activities. It is important that the fund's supporters do not skip over the process steps of establishing the fund's purpose and need and identifying the top-loss investor. These activities can demonstrate that the community understands the fundamentals necessary to start a fund and make the fund attractive to prospective investors.

The TOAH and Denver TOD funds illustrate alternative models for establishing a fund, including various ways to organize capital and attract investors. However, in both cases, perhaps the biggest challenge in starting a structured fund for TOD across a region is a source of top-loss investment that can support loans across multiple jurisdictions. The Denver TOD Fund was not able to accomplish this goal, but the TOAH Fund was able to because the fund's supporters found a willing partner in MTC.

The Environmental Protection Agency (EPA) selected four applicants to its Smart Growth Implementation Assistance (SGIA) program who had all requested assistance with financing infrastructure related to transit-oriented development (TOD)[1] and combined assistance to these communities into one project. The project resulted in this report. In addition, the consultant team provided each selected applicant with a memo describing how some of the infrastructure financing tools might be applied in each community. This appendix presents the information from those memos.

The four applicants selected through the SGIA program are:

- Cobb County and the Cumberland Community Improvement District, Atlanta, Georgia, who asked for help with identifying potential funding sources for large, high-capacity transit projects, land acquisition around potential transit stations, and infrastructure to facilitate TOD.

- South Suburban Mayors and Managers Association (Chicago area), Illinois, which asked for help identifying funding and financing tools to facilitate TOD and evaluating the potential role of a structured fund and land bank in TOD.

- Utah Transit Authority (UTA), Salt Lake City, and Sandy City, Utah, which asked for assistance in evaluating financing mechanisms for non-revenue generating infrastructure and developing a funding strategy for TOD infrastructure, specifically structured parking, in one station area.

- Wheat Ridge Colorado, which asked for help in developing a funding strategy for TOD infrastructure in a station area.

---

[1] TOD can be defined as development within a quarter- to a half-mile of a transit station that generates ridership for the transit system, lowers peoples' transportation costs, and increases housing and transportation choices. For a more detailed description of TOD, please see the report's introduction in Chapter I.

A. COBB COUNTY AND THE CUMBERLAND COMMUNITY IMPROVEMENT DISTRICT

LOCAL CONTEXT

The Cumberland Community Improvement District (Cumberland CID) was created in 1987 through a resolution by the Cobb County Commission and the consent of commercial property owners in the Cumberland area. It is a 5.5-square-mile area in southeast Cobb County that includes the intersections of I-75, I-285, and U.S. Highway 41 (see Exhibit D-1). Commercial property owners in the area fund the Cumberland CID by paying an additional five mils (5/1000) of property taxes annually. Cobb County collects the taxes and distributes those funds to the CID. In 2010, the Cumberland CID collected $5.3 million in net commercial property taxes, about 90 percent of which was dedicated to transportation infrastructure, programs, and planning, with the balance for operating overhead.

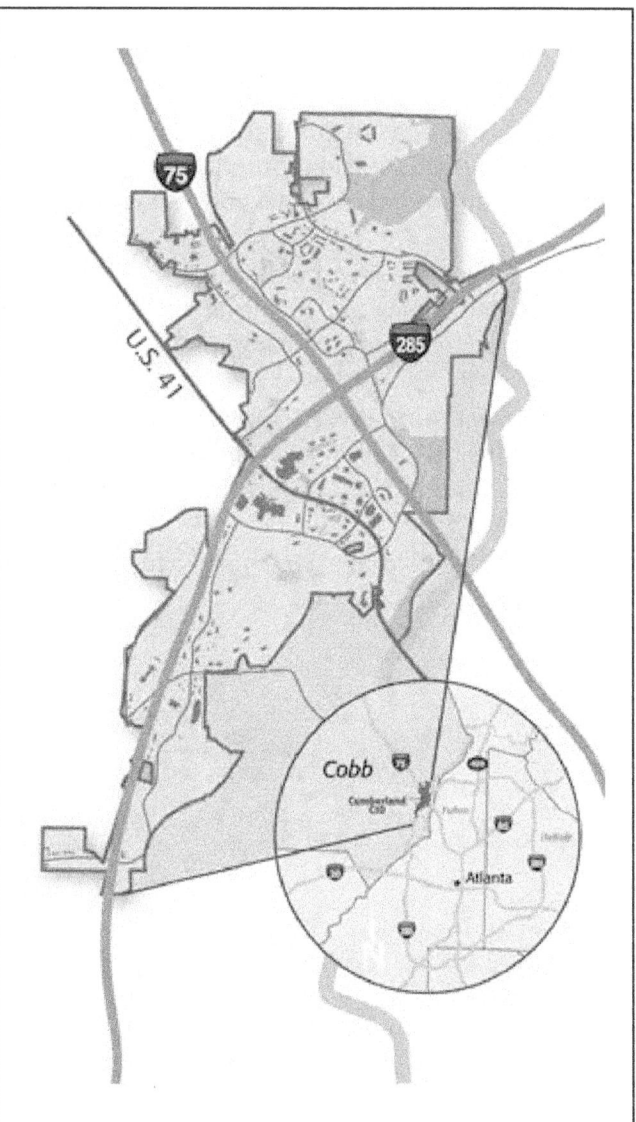

The Cumberland area has more than 17.5 million square feet of office space and 3.5 million square feet of retail in the CID and more in the surrounding areas. Cumberland is one of Georgia's largest employment centers, home to more than 65,500 full time jobs, or 1.7 percent of Georgia's total jobs, with an economic impact of more than 5 percent of the state's total economy. By 2020, between 2 to 3 million more square feet of office space will likely be needed to accommodate anticipated new jobs.[2]

TOD PROJECTS AND TOD INFRASTRUCTURE NEEDS

STATUS OF TOD PROJECTS

Approximately 40 percent of Cobb County commuters work in Cobb County, and another 29 percent work in Fulton County. Approximately one-third of these commuters

Exhibit D-1. Map of Cumberland CID.
Source: Cumberland CID.

---

[2] Cumberland Community Improvement District. *Annual Report*. 2010. http://www.cumberlandcid.org/files/media/documents/2010-ccid-annual-report.pdf.

use the U.S. 41 corridor. Although Cumberland CID is a regional employment hub, transit services connecting the county to Atlanta are limited.

The Metropolitan Atlanta Rapid Transit Authority (MARTA) is the principal transit system in the Atlanta metropolitan area. MARTA operates a network of bus routes linked to a rail system consisting of 48 miles of track and 38 train stations. MARTA operates almost exclusively in Fulton and DeKalb counties, with bus service to two destinations in Cobb County (the Cumberland Transfer Center within the CID and Six Flags Over Georgia). In addition, Cobb County offers local and express bus service to points around the county and to the MARTA Arts Center station. The express bus route to the Arts Center station is one of the highest volume routes in the MARTA system and has a fare box recovery[3] of 47 percent, above the national average.

To improve transit services, Cobb County is exploring high-capacity transit alternatives that could connect its employment centers to Atlanta. Among the options considered is a bus rapid transit (BRT) system and/or a potential light rail line.[4] U.S. 41 is considered to be a potential transit corridor, running approximately 14 miles from midtown Atlanta through Cumberland and connecting to Town Center Area in north Cobb County.

The Transportation Investment Act of 2010 (TIA 2010) created a legal mechanism under which Georgia regions can impose a 1-cent sales tax to fund needed transportation improvements. No counties or municipalities are permitted to be exempt from the tax if it is approved by a simple majority of voters across the entire region. The tax is valid for up to 10 years but could be extended if approved through subsequent voter referendums.

In July 2012, voters in 10 metropolitan Atlanta counties voted against a 1-cent sales tax to pay for $7.2 billion worth of road and transit projects of regional significance. Cobb County and its municipalities would have received approximately $1 billion over 10 years, primarily for roadway improvements and transit options. The construction of a BRT system from midtown Atlanta to north Cobb County was positioned for $689 million in TIA 2010 investments. Stations, parking, and right-of-way acquisition, along with maintenance and operating costs, were included in the funding, but additional resources would have been needed to develop the sites around the stations. To explore other transit alternatives in the corridor, Cobb County is partnering with FTA on an alternatives analysis to research light rail and other potential transit solutions that could upgrade the local BRT investment to a light rail project if deemed appropriate and future federal funds are secured.

INFRASTRUCTURE NEEDS

No specific projects related to TOD were identified by Cobb County and the CID at the time of this analysis since these entities were in the early stages of planning the transit mode and the corridor alignment. The tasks to be completed over 2012-2013 include developing a draft Environmental Impact

---

[3] The fare box recovery is the percent of operating expenses that are covered by passenger fares.

[4] Cobb County, Board of Commissioners Work Session. May 25, 2010
http://dot.cobbcountyga.gov/Planning_Studies/VISIONS/BOC_PPT_May_2010.pdf.

Statement and determining funding sources for construction and operation.[5] There are four proposed station locations; however, the number could change depending on a future engineering study.

The Cumberland CID and Cobb County identified the following infrastructure needs for supporting TOD:

- High-capacity transit connecting to the regional transit system.

- Secure land and/or facilitate development around potential transit stations to encourage TOD.

MOST APPLICABLE TOOLS AND STRATEGIES

Under Cobb County's legal and economic framework, the following tools seem to be most applicable at this point:

- Public-private partnerships (most appropriate for large projects).

- Revenue bonds (appropriate for smaller projects).

- Land banking (appropriate for smaller projects).

- Structured funds (appropriate for smaller projects).

TOD Infrastructure	Public Private Partnership	Revenue Bond	Land Bank	Structured Fund
Large-scale development (land acquisition, transit facilities, real estate)	X			
Land acquisition	X	X	X	X
Small projects		X		X

Exhibit D-2. Summary of TOD infrastructure financing options for Cobb County.

PUBLIC- PRIVATE PARTNERSHIPS

Public-private partnerships[6] can bring private equity to expedite infrastructure projects because the community does not need to wait until sufficient public funds are available. Public-private partnerships are generally suitable only for large-scale projects because they involve high implementation costs. The construction of the light rail or the BRT from Atlanta to Cobb County could entail at least $600 million in infrastructure investment and thus is a good candidate for establishing a public-private partnership. A public-private partnership can be structured in many ways depending on the needs and funds of the public sector. Some examples of public-private partnership arrangements include:

- **Limited private participation:** The private sector could be responsible for financing, building, and maintaining the stations or other rail components only, while the public sector could be responsible

---

[5] Northwest Corridor Light Rail Transit System. *Transit Implementation Study.* 2010. http://dot.cobbcountyga.gov/Planning_Studies/VISIONS/Transit_implementation_Study_Aug2010.pdf.

[6] More detail on public-private partnerships is in Appendix B, Section D-1.

for financing other TOD elements such as parks, streetscape improvements, streets, bike and/or pedestrian improvements, and structured parking. The private sector could be paid back for its investment in installments called availability payments from the general fund. (See the Long Beach Courthouse example described in Appendix B, Section D-1). One advantage of having the private sector maintain the new infrastructure is that the infrastructure tends to last longer and provide better levels of service because it receives adequate maintenance; the private sector can be subject to payment deductions if it does not maintain the infrastructure.

- **Extensive private participation:** The private sector could be responsible for financing, building, and maintaining the transit stations and other TOD elements such as streetscape improvements, parking, [7] and bike paths. The private sector could also develop the land around the station into commercial and residential uses. Property taxes from new development could be used to pay back the private sector for the public infrastructure, as in the New Quincy Center example (described in Chapter III, Section E of this report). In Quincy, the private sector will bear the construction, design, and financial risks of developing TOD infrastructure. The city will reimburse the developer through taxes captured by a special assessment district on new development. However, the city will proceed with reimbursements only at a certain occupancy threshold, ensuring that it will receive sufficient income from property taxes from new development to reimburse the developer. Georgia's legislation allows the creation of a special assessment to revitalize business districts. If some of the areas where the transit line will be built qualify as business districts in decay, then a special assessment, if approved by voters, could be feasible. Another possibility to increase potential sources of revenue to pay back the private sector would be to expand the size of the existing CID, assuming that those additional properties would benefit from TOD.

Other funding sources could be explored to pay back the private sector, including general funds from the cities that will benefit from the BRT and/or light rail; CID revenue from Cumberland and Town Center (another CID); the creation of new CIDs along the corridor, if feasible; or a combination of general funds and CIDs revenue.

STEPS/TIMING

Georgia has enabling legislation to support the creation of public-private partnerships; however, before Cobb County and/or the Cumberland CID decide to use a public-private partnership to deliver a project, they need to do a thorough analysis of the project's legal, technical, and financial feasibility. This analysis requires:

- Assessing the project's funding gaps. As soon as the funding gaps for the high-capacity transit project have been identified, Cobb County and the Cumberland CID could explore the possibility of implementing a public-private partnership, which is a time-consuming task that could take from two to four years, depending on the complexity of the project.

---

[7] The structured parking is not assumed to generate revenue because parking in Cobb County is currently free; paid parking would not be able to compete with free parking. For a parking structure to be able to generate revenue from parking fees, the county or the city would need to gradually modify parking policies from free to paid parking.

- Identifying which project or project components are suitable to be delivered under a public-private partnership framework based on existing legislation. Elements could include all transit components including the stations, tracks, and vehicles, or only the stations and tracks.

- Completing environmental reviews.

- Engaging advisors. The project sponsor, in this case Cobb County, would generally issue a request for proposals to select technical, legal, and financial advisors to guide it through the transaction from project screening to the preparation of bid documents (contracts and legal, financial, and technical requirements), evaluation of the proposals, and negotiations with the preferred bidder.

- Performing a "value for money" analysis, which consists of comparing the value of a project to the public sector under two different procurement methods—traditional public procurement (design-bid-build) versus public-private partnership—due to differences in financing costs and risk valuation.

REVENUE BONDS

Revenue bonds[8] are a type of municipal bond that is secured by a specific revenue stream. Revenue bonds can be issued by cities, counties, and, in some states, by special districts to finance improvements for a revenue-producing enterprise. Revenue bonds are repaid solely from the revenues generated by the financed facility. The Georgia Constitution (Article IX, Section VII) allows CIDs to issue bonded debt, but such debt cannot be considered an obligation of the state or any unit of government other than the CID.

One of the benefits of issuing debt is the ability to expedite project delivery as opposed to a pay-as-you-go approach. Currently the Cumberland CID does not leverage on the revenue generated by the district to issue debt. However, the CID could use commercial property tax revenue for bond repayment and support small-scale projects such as utilities or parks.

STEPS/TIMING

The Cumberland CID could hire an advisor to determine the feasibility of issuing debt supported by its revenue or do a high-level analysis in-house. The revenue bond could be used in the short and medium terms to help address the district's funding needs. To issue bonds, the following steps are typically necessary:

- Identify funding needs.

- Assess the potential for grouping projects. Depending on the size of the funding needs, small projects could be grouped to reduce the transaction costs of issuing a bond.

- Assess the Cumberland CID's debt capacity based on existing commitments, revenues, and the CID's remaining life, since no CID can incur financial obligations, including bonding, beyond its existing life. The Cumberland CID's life terminates every 6 years unless the CID membership votes to extend for

---

[8] More information on revenue bonds is in Appendix B, Section B-3.

another 6 years. The CID's current life is through April 2018. The term of the CID will dictate the maximum term of the bond that could be issued and the potential revenue to repay the debt.

- Involve a specialized group of advisors to assess the feasibility of issuing a bond, including a financial advisor, an underwriter, bond counsel, a rating agency, and insurers.

LAND BANKING

Most land banks[9] are public authorities created to acquire vacant or foreclosed properties to stabilize neighborhoods or create affordable housing. There is no precedent of land banks acquiring properties or land for TOD, not necessarily because states' legislation precludes it, but likely because land acquisition and management is a complex task that requires specialized staff in finance, real estate, marketing, and law. However, the current weakness of the real estate market in most metropolitan areas around the country, including the Atlanta metropolitan area, offers agencies an opportunity to assemble land at low cost to support TOD, which could then be developed into mixed-use development. The revenue generated from leasing or selling the land adjacent to the stations could generate profits that could be used to repay the loan for land acquisition. One critical aspect that Cobb County and the Cumberland CID need to consider is the cost associated with holding the land (property taxes) until the timing is appropriate to sell or lease to potential developers. There are areas in Cobb County near the potential station areas with possibilities for higher-density development, given that the land is either underdeveloped or redevelopable.

STEPS/TIMING

Implementing a land bank would require the following steps:

- Identify land acquisition needs.

- Decide the purpose of the land bank, the types of projects to fund (e.g., transit or real estate) and the type of clients that will have access to capital. Would land acquisition be a one-time endeavor, or would it be part of the county's long-term policy? The answer to this question will determine the resources needed to support land acquisition. Other counties or cities in the Atlanta metropolitan area might share similar needs for land acquisition, and coordination with them could build the case for a land bank and make it easier to access state or federal resources.

- Establish the land bank's institutional framework. Land banks are typically structured as revolving loan funds, which are allowed under Georgia law. An independent entity would need to be created to acquire and manage land with a staff for property acquisition and disposition, due diligence, pro forma budgeting, leasing, property management, asset management, financial reporting, lease administration, and accounting.

- Identify the land bank's funding resources. The land bank could be funded through initial capitalization in the form of "seed money," an ongoing stream of dedicated revenue, or both. The capitalization could come from an initial appropriation from Cobb County's or Georgia's general

---

[9] More information on land banking is in Chapter IV, Section C.

funds, grants from corporations or foundations, borrowing of capital funds, or the proceeds of a one-time asset sale. The ongoing revenue stream could be a dedicated portion of an existing or new tax. Cobb County could also consider dedicating a portion of its existing property tax to provide the initial capitalization or seed money.

STRUCTURED FUNDS

A structured fund is a pool of contributions from multiple investors with different risk and return parameters. Capital contributions come from public and private for-profit and nonprofit sources, resulting in risk diversification and a reduction in the cost of capital. Structured funds are typically targeted to land and/or property acquisition and are linked to a specific social purpose.[10]

STEPS/TIMING

The creation of a structured fund in Cobb County would involve the following steps:

- Identify the purpose of the fund. For example, the fund could be dedicated to land acquisition around potential transit stations, which the Cumberland CID has cited as a need. If the land acquired around stations is targeted for future real estate development, such as housing and commercial space, the revenue generated through land sales or leases could be used to repay the loans from the fund.

- Identify capital sources for the fund. Sources from multiple investors include grants from public or private institutions willing to take higher risk (without return expectations) and private sources expecting returns on investments.

- Determine the fund's institutional framework. The fund would need to be a separate entity but, as with some revolving loan funds, it could be housed in another state agency, such as the state treasury department, the Georgia Transportation Infrastructure Bank, or an independent entity.

Establishing agreements with potential investors is a time-consuming process. Work to establish a structured fund could start as soon as funding needs have been identified.

NEXT STEPS

- **Step 1: Define the Project.** Refine the high-capacity transit project description, initiate its environmental impact statement, and determine what the complementary TOD infrastructure elements are (e.g., streetscape, parks, bike paths). Further definition of the project, including the number of stations and right-of-way needs will translate into more detailed capital and operating cost estimates, which can be used as the baseline to assess investment needs.

- **Step 2: Assess Funding Gap Scenarios.** Cobb County could assess the high-capacity transit and TOD elements funding gap scenarios. A conservative approach would consider as part of the funding mechanisms only resources that are likely to be committed to the project.

---

[10] Structured funds are described in more detail in Appendix C. Fundamentals of Structured Funds.

- **Step 3: Screen Tools Based on Funding Gap.** Once the funding gap for each project has been determined, Cobb County and the CID could screen potential tools in more detail to assess both the funding that could be generated and the pros and cons of each tool based on the community's specific project needs. Because voters rejected the 1-cent sales tax, the funding gap will be large, and thus tools that are more suitable for large capital projects, such as public-private partnership, combined with other tools like land banking have the potential to shorten capital needs.

B. SOUTH SUBURBAN MAYORS AND MANAGERS ASSOCIATION

LOCAL CONTEXT

SSMMA covers an area including 42 municipalities in the suburbs south of Chicago. The area includes 33 transit stations on four Metra[11] commuter lines and a proposed new Metra rail line that will serve nine additional stations. Real estate values of south suburban station areas are generally lower than other suburban station areas in the Chicago region.

EPA and SSMMA asked the consultant team to focus on the Vermont Street Station area in Blue Island, since it is representative of the opportunities and challenges faced by communities throughout the SSMMA area. Blue Island is a city about 20 miles south of Chicago with a diverse population of just over 25,000. Blue Island's economy has historically been based on industrial uses, but the city suffered significant job losses as manufacturing has declined over the last several decades. Three Metra commuter rail lines converge in Blue Island's Vermont Street Station area, which includes two stations that face each other across Vermont Street. The stations currently have 90 trains a day to and from the Chicago Loop; only two other Chicago suburbs have comparable levels of commuter rail service to the Loop.

The Vermont Street Stations are about a quarter-mile from Blue Island's Main Street District and a quarter-mile from MetroSouth Medical Center, a regional hospital with 1,300 employees, but both destinations are up a fairly steep hill from the stations, and the area lacks good pedestrian access (see Exhibit D-3).

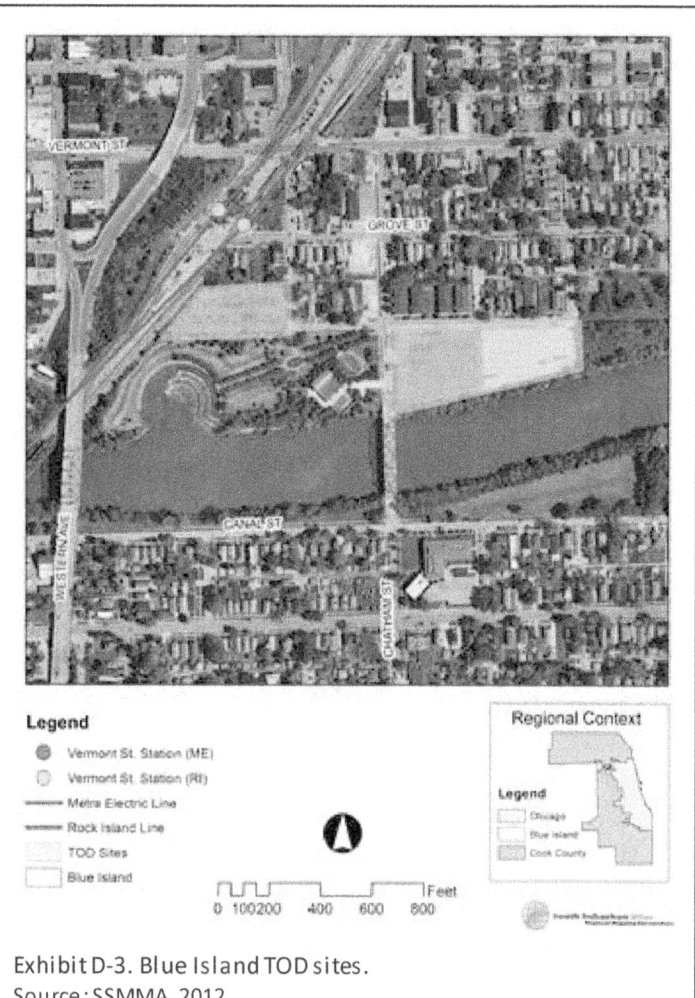

Exhibit D-3. Blue Island TOD sites.
Source: SSMMA, 2012.

---

[11] Metra is a commuter rail agency with 11 lines running from Chicago's downtown to 241 stations in Cook, DuPage, Will, Lake, Kane, and McHenry counties.

TOD PROJECTS AND TOD INFRASTRUCTURE NEEDS

STATUS OF TOD PROJECTS

Blue Island's TOD plan, which was developed in 2006 and updated in 2009 and covers several TOD areas, including the Vermont Street Station area, calls for:

- Development of condominiums or mixed-income rental housing on the vacant and currently industrial land near the station.

- A linear park along the waterfront with adjacent housing.

- Improved pedestrian access between the stations, the Main Street District, and the hospital.

- Streetscape improvements in the Main Street District.

As of early 2012, the city's efforts to implement the TOD plan include:

- Performed Phase I assessments of potential brownfield sites that subsequently proved not to need remediation.

- Made some minor streetscape improvements in the Main Street District.

- Established a TOD-supportive zoning district for the TOD and Main Street areas.

- Formed a TIF district that includes the hospital but not the Vermont Street Station area.

In addition, the city has sought but not received state and federal funding for pedestrian access improvements between the stations, the Main Street District, and the hospital.

The Vermont Street Station area includes the following land uses:

- Twelve acres of industrial uses owned by four companies. The city would like to relocate the industrial uses outside of the Vermont Street Station area.

- Eight acres of surface parking lots serving Metra, including a four-acre parking lot overlooking the Cal-Sag Channel that Metra has identified as unneeded. According to SSMMA, Metra has stated that if the parking area is redeveloped, it will not require replacement parking. The lot is owned by the Metropolitan Water Reclamation District. SSMMA reports that as of early 2012 several developers have expressed interest in acquiring the site.

- Six acres of park land along the Cal-Sag Channel.

- Twelve acres of vacant and undeveloped land along the Cal-Sag Channel.

- Single-family homes and a mix of multifamily housing and retail along Vermont Street.

INFRASTRUCTURE NEEDS

SSMMA identified structured parking as a primary need for TOD infrastructure in Blue Island. Although one Metra parking lot could be developed without building replacement parking, a second Metra parking lot has also been identified for redevelopment, and redeveloping both sites would require adding a parking structure. Other infrastructure needs that SSMMA identified include better connections to regional bike paths and between the station area and the hospital.

Although costs for replacement parking or other infrastructure needed in the station area were not available at the time of this analysis, SSMMA provided a preliminary financial analysis showing the estimated costs and revenues of two conceptual development scenarios. The financial analysis is intended to provide Blue Island "with a very broad 'bird's eye' view as to whether the project is at all feasible and therefore it may move toward being made available to the marketplace for review."[12] Although it was still subject to change at the time this report was written, the preliminary financial analysis indicated that for the two conceptual development scenarios, total costs exceed total revenues, and the projects would therefore not be financially feasible without public investment.

MOST APPLICABLE TOOLS AND STRATEGIES

The following strategies for preparing for development could be considered:

- Expansion of the existing TIF district.

- Creation and/or expansion of relationships with anchor institutions.

- Future evaluation of a structured fund or a land bank.

TOD Infrastructure	Tax Increment Financing	Anchor Institutions	Structured Fund or Land Bank
Bike and pedestrian improvements	X	X	
Land acquisition	X		X
Street improvements	X	X	
Streetscape improvements	X	X	
Structured parking	X		

Exhibit D-4. Summary of TOD infrastructure financing options for Cobb County.

TAX INCREMENT FINANCING

TIF[13] allows the public sector to "capture" growth in property tax resulting from new development and increasing property values. In Illinois, communities can use TIF for individual projects or within a district. A TIF district can be a powerful value capture tool because it capitalizes on increases in property values,

---

[12] Business Districts, Inc., "Blue Island Site Analysis," January 12, 2012.

[13] More information on TIF is in Appendix B, Section E-3.

including the value of new development, in an entire district, making it possible to spread the costs of infrastructure across a wider base. The city has already established a TIF district, but it does not cover the station area. The TIF could be expanded to include the Vermont Street Station area.

STEPS/TIMING

Blue Island could begin the process of expanding the TIF district in the short term.

ANCHOR INSTITUTIONS

Blue Island could work with the existing anchor institution, MetroSouth, to better understand the hospital's transportation and employee housing needs that might overlap with the infrastructure needs in the Vermont Street Station area.[14] There are opportunities for the city and the hospital to meet multiple goals together.

The city could cooperate with the hospital to assess if MetroSouth could contribute to the efforts to improve connections between the station area, the hospital, and the Main Street District. This assessment could evaluate how hospital employees are currently getting to work and if more employees would use Metra if a shuttle or better pedestrian connections from the station were available. The city could also determine if the hospital has expansion plans or needs additional parking. The hospital might be able to better meet its own transportation needs in cooperation with the city in a way that would help pay for the connections between the station, the hospital, and the Main Street District. For example, if the hospital could reduce the number of new parking spaces needed in a parking garage due to better transit and pedestrian access, it might be willing to contribute money to pay for the pedestrian connections. Similarly, the hospital might help to pay for regional bike paths to encourage its employees' healthy living. The city could also assess hospital employees' need for housing to determine if they could generate demand for housing in the station area.

STEPS/TIMING

Blue Island could begin discussions with MetroSouth and any additional anchor institutions in the short term.

STRUCTURED FUND OR LAND BANK

SSMMA is interested in how a structured fund or land bank could help provide TOD infrastructure. As described in, a structured fund is a loan fund that pools money from different investors with varying risk and return profiles. Structured funds have a dedicated purpose, which is clearly defined prior to forming the fund, and are managed by professionals with fund formation and loan underwriting experience. Recently there has been increasing interest in using structured funds as a property acquisition tool to support affordable housing development, specifically near transit. SSMMA is in the process of forming a structured fund. Since forming a structured fund will likely take several years, there is time to focus on other activities before the fund is ready to operate. Once the form and purpose of the SSMMA

---

[14] More information on working with anchor institutions for financing TOD infrastructure is in Chapter IV, Section A.

structured fund are settled, the Blue Island project could be assessed to determine if it is eligible under the fund's selection criteria.

A land bank[15] could acquire land in the area that is not yet ready for development and hold it until appropriate development is possible. However, to attract private investment capital a land banking authority would likely need to transfer the land to a developer at no cost or a very low cost. Because the sites that will likely be the first locations for development are already held by government agencies, land banking is a low priority for the Vermont Street Station area.

STEPS/TIMING

A structured fund or land bank is a mid- to long-term action.

NEXT STEPS

The city of Blue Island will be the lead entity in implementing TOD projects, but because SSMMA received this technical assistance, some of the next steps are directed to SSMMA. In many places in the implementation process, the city will be a facilitator, not a financer, as it works with MetroSouth, Metra, and potential developers, and SSMMA can help the city with the process.

- **Step 1: Conduct Analyses.** To better understand the development likely to occur in the station area, the city could prepare a market and affordability analysis to estimate the demand for specific land uses. This analysis, including evaluation of the affordability of included housing unit types, could be started immediately. A key question for the analysis is if local incomes can pay the rents or sale prices necessary to support the building types envisioned for the area. If not, the affordability analysis could estimate how long it might take to overcome an affordability gap. The market and affordability analysis could specifically consider the potential for demand for housing from the hospital or other anchor institutions.

  This analysis could also include an opportunity site analysis that determines which parcels have the most potential for redevelopment. That analysis could help SSMMA and the city better focus their efforts. The analysis would also allow SSMMA and the city to assess the need to relocate existing industrial uses in the station area. The analysis could test the development plan to see what the short- to mid-term demand is for housing and if that demand can be met without relocating the existing industrial uses. An additional financial analysis could assess the feasibility of projected building types under current and projected market conditions and policies. Conducting a targeted market and affordability analysis would provide SSMMA and the city with specific information on the potential for development and the policy changes and market conditions required for successful development.

- **Step 2: Create and Strengthen Partnerships.** SSMMA could work with the city to create and strengthen partnerships with anchor institutions. SSMMA and the city could meet with representatives of MetroSouth to discuss the hospital's plans for expansion and transportation needs. These discussions could assess the potential for partnering with MetroSouth on better

---

[15] More information on land banks is in Chapter IV, Section C.

connections between the station area, the hospital, and the Main Street District on locating hospital-employee housing in the station area. SSMMA and the city could also reach out to the property owners in the station area to assess their plans for expansion, relocation, or redevelopment. SSMMA could help build partnerships by reaching out to and educating the community. The partnership efforts could be started immediately.

- **Step 3: Build Value.** SSMMA and the city could use the results of the market and feasibility analyses from Step 1 to identify the most appropriate location(s) to begin redevelopment. The most appropriate location to make public investments could be a subarea or a project that needs a smaller investment from the city. Smaller projects that incrementally increase the value of the area can build momentum and value that can be leveraged to help pay for larger projects in the future. As property values increase in the area, the TIF district will have more resources to invest in larger projects. In real estate development, the most expensive projects often take the longest time, so a TIF district might not be able to reap the benefits immediately. Therefore, the larger projects can (or must) sometimes wait as value is built around them.

  To help build momentum and value, the city can undertake small improvement projects in a targeted area, such as streetscape improvements on one or a few blocks in the station area, and encourage small rehabilitation projects, such as façade improvements, that signal the market that investments are being made in the area. These types of incremental steps can help the city "set the table" for private development. Showing potential developers a successful, completed project can help to persuade them to invest in the area.

  This strategy of building value could also reduce the need for public funding to help relocate industrial uses in the area. Identifying a subarea where a critical mass of successful developments can be achieved could begin to build momentum and value that encourage other property owners and land uses to move on their own in the future.

  This step could also include expanding the TIF district to include the Vermont Street Station area. Expanding the TIF district would allow the city to capture the incremental increases in value and apply the revenue to projects in the station area. The expansion of the TIF district could be undertaken immediately. Other efforts under this step could begin as information becomes available from the market and feasibility analysis and from discussions with key partners that enable the city to determine a target subarea.

- **Step 4: Leverage Opportunities.** In 2010, SSMMA was awarded a $2.3-million Sustainable Communities Challenge Grant from HUD to implement its Green TIME Zone strategy[16] and take advantage of existing and planned transit and related housing and economic development opportunities. SSMMA could explore how to leverage its HUD grant to help identify and obtain additional resources (e.g., philanthropic grants or state and federal funding) for neighborhood-level improvements (e.g., pedestrian and bicycle improvements between the stations, Main Street

---

[16] The Green TIME Zone strategy is designed to help older communities use their existing rail infrastructure and manufacturing capacity to help create better neighborhoods and jobs and improve the environment. See: Center for Neighborhood Technology. *Chicago Southland's Green TIME Zone*. 2010. http://www.cnt.org/repository/GTZ.pdf.

District, and MetroSouth; streetscape improvements; and façade improvements) that can help to build value in the station area.

C. UTAH TRANSIT AUTHORITY (UTA), SALT LAKE CITY, AND SANDY CITY, UTAH

LOCAL CONTEXT

UTA is the public transit provider for the most populated counties in Utah, including Salt Lake, Weber, Davis, and Utah counties. These four counties make up approximately 70 percent of Utah's population and are in the mountain-surrounded corridor of the Wasatch Front, which extends approximately 100 miles north to south.

In 2010, the Utah legislature passed Senate Bill 272, which authorized UTA to enter into agreements with developers as a limited partner on up to five sites owned by UTA. The purpose of this authorization is to increase transit ridership by supporting TOD and increasing UTA's self-reliant operating funds. Under SB 272, UTA can contribute portions of land it owns around transit stations to a developer's project in exchange for a say in how to develop the land and a share of the profits.

Of the UTA-owned sites throughout the region 27 could potentially support TOD because they are designated as excess land, meaning they are no longer needed solely for a transit purpose. UTA has created a TOD department to ensure that it makes good use of the powers granted under SB272.

UTA works closely with the region's MPOs and Envision Utah, a nonprofit partnership that facilitates community planning throughout the Wasatch Front, to create visions and strategies for developing key sites owned by UTA and along UTA transit corridors. A driver of these organizations' efforts is the "3 percent strategy," developed by Envision Utah under which the region intends to locate a third of its growth on just 3 percent of the region's developable land linked by a world-class transit system.[17]

Together, these organizations have identified six demonstration sites that are prime for development that could catalyze other TOD in the region. EPA and UTA asked the consultant team to focus on one of those sites, the Sandy Civic Center TRAX Station site. UTA owns approximately 48 acres in the area. Much of the UTA-owned land is currently used as surface parking or is undeveloped. The station area is within walking distance of the Rio Tinto Stadium (a major league soccer stadium), South Towne Exposition Center (a conference and event center), Sandy City Hall, Sandy Business Park, and South Towne Center (a regional shopping mall).

TOD PROJECTS AND TOD INFRASTRUCTURE NEEDS

STATUS OF TOD PROJECTS

In 2010, UTA issued a request for qualifications and financial proposals for the development of the 48 acres it owns around the Sandy Civic Center TRAX Station site. The objective of the request was to build a high-density, mixed-use community that will increase ridership, generate long-term revenue, and integrate the station and development in a manner that will encourage and support transit use. UTA has also planned direct bicycle, pedestrian, and horse-trail connections between the site and the Dimple Dell nature preserve in the Wasatch Mountain Range. The 48-acre site is broken into two parts running along

---

[17] Envision Utah. *The 3% Strategy*. undated. http://envisionutah.org/ThreePercentStrategy.pdf.

the rail line from north to south. A developer has developed a master plan for part of the 48-acre site that includes 1,185 residential units, 300,000 square feet of office, and 59,000 square feet of retail (Exhibit D-5). The plan also includes a village square, a transit plaza, and trails connecting to existing open space. The developer estimates a seven- to 10-year buildout of the planned development with Phase 1 beginning in the fall of 2012. Phase 1 development includes 168 residential units, 30,000 square

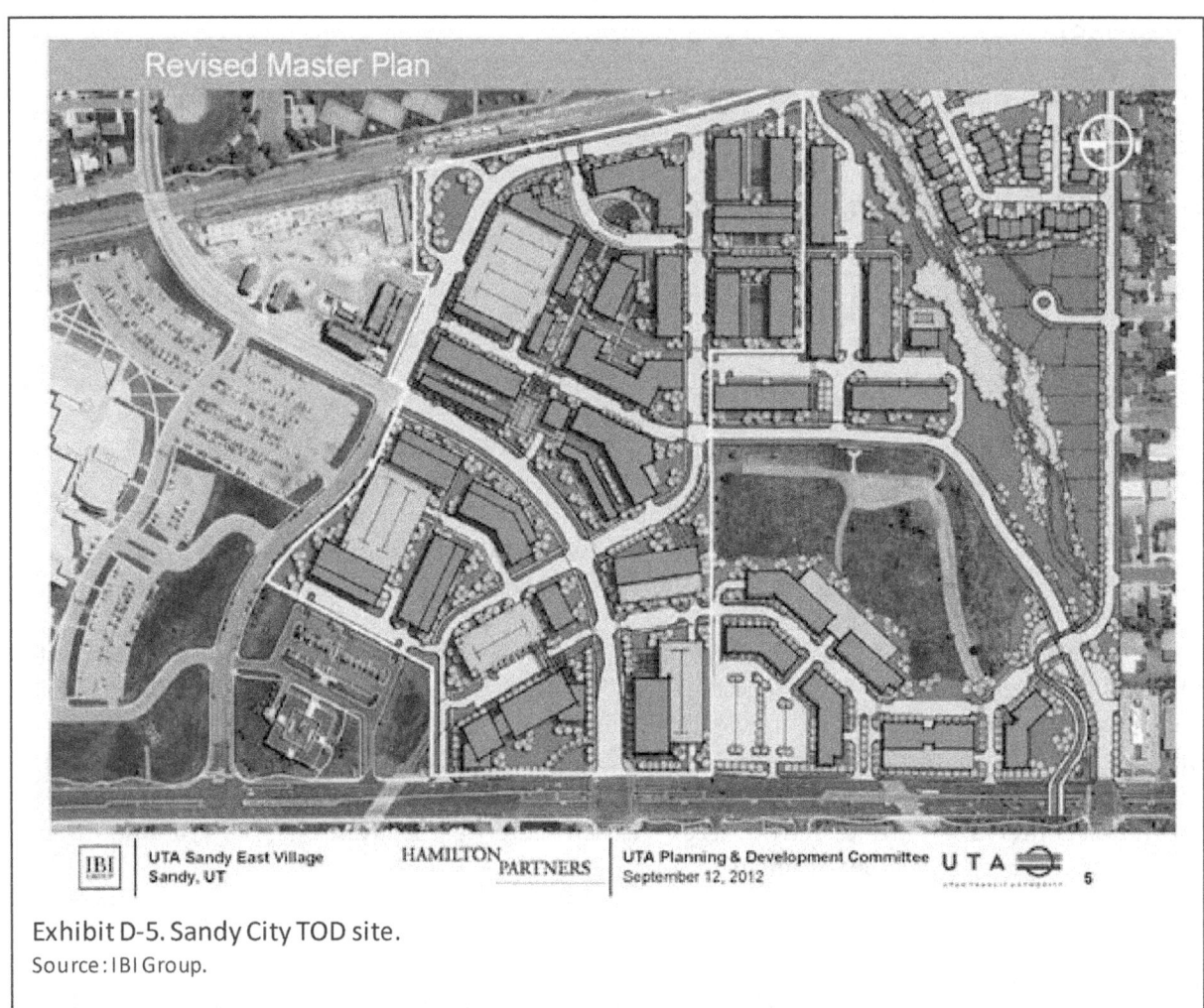

Exhibit D-5. Sandy City TOD site.
Source: IBI Group.

feet of retail, 570 surface parking spaces, and 340 structured parking spaces.

INFRASTRUCTURE NEEDS

Among the infrastructure needs for the Sandy Civic Center Station area, the city and UTA have identified stormwater facilities and other utilities in addition to neighborhood amenities such as streetscape improvements and parks. However, structured parking was identified as the primary need for this technical assistance because UTA has found it to be the most difficult to fund or finance. About 1,200 surface parking spaces are currently located in the station area. However, because of new stations opening in the transit system, the total number of spaces needed to serve the station is expected to drop in the future. UTA estimates that an 850-stall parking structure, costing about $16 million, will be needed. UTA has identified $2 million in funding for the parking structure, leaving an estimated $14 million unfunded.

MOST APPLICABLE TOOLS AND STRATEGIES

Sandy City and UTA might be able to address the station's parking needs with a strategy that combines:

- Excess property disposal.

- Joint development.

- Shared parking.

TOD Infrastructure	Excess Property Disposal	Joint Development	Shared Parking
Parks		X	
Stormwater facilities		X	
Streetscape improvements		X	
Structured parking	X	X	X

Exhibit D-6. Summary of TOD infrastructure financing options for UTA and Sandy City.

DISPOSAL OF EXCESS PROPERTY TO SUPPORT TOD

Of the UTA-owned 48 acres, 29 acres are adjacent to the TRAX station, while another 19-acre parcel runs along the TRAX line to the south. While the development program has not been set for the site, the most recent concept available to the consultant team includes only the 29 acres closest to the station. UTA could sell the 19-acre parcel and use the revenue to fund infrastructure in the station area. To ensure that any development on the 19-acre parcel is compatible with UTA's and Sandy City's TOD goals, UTA can pursue a deed restriction on the sale of the parcel that requires a certain density.

Because UTA acquired the property with the participation of the FTA, this disposition strategy will also likely require FTA's participation.

STEPS/TIMING

Exploring the potential to sell the 19-acre parcel could begin immediately. However, UTA would need to consider many factors in determining if a sale should go forward, including the appraised value of the parcel, current market considerations, and any relevant discussions with FTA and potential developers.

JOINT DEVELOPMENT

Joint development on publicly owned land near the transit station could help fund needed TOD infrastructure (see Section 19. Since FTA participated in the purchase of the 19-acre parcel, FTA's joint development and excess property disposition policies must be considered. Under FTA policy, a joint development project can include: commercial and residential development that is physically or functionally related to public transportation projects; pedestrian and bicycle access to a public transportation facility; construction, renovation, and improvement of intercity bus and intercity rail stations and terminals; and renovation and improvement of historic transportation facilities.[18] Further, to be eligible for federal funding, a joint development project must:

- Enhance economic development or incorporate private investment.

- Enhance the effectiveness of a public transportation project or establish new or enhanced coordination between public transportation and other transportation.

- Provide a fair share of revenue to be used for public transportation.[19]

Under FTA's joint development laws, activities that can be funded from property disposal proceeds include:

- Real estate acquisition.

- Demolition.

- Project development activities.

- Site preparation.

- Building foundations.

- Parking.

- Transportation-related furniture, fixtures, and equipment.

- Utilities.

- Walkways.[20]

Based on the list of eligible activities, UTA could sell the property and use the proceeds to help pay for the TOD infrastructure needed at Sandy Civic Center Station, such as a parking garage. FTA also notes that the proceeds from disposal actions can be used "to reduce the gross project costs of another eligible capital project. This may include approved joint development projects. Note that a transfer of real property meeting the tests for joint development is not a disposition, and the proceeds are deemed program income."[21]

---

[18] DOT. "Notice of Final Agency Guidance on the Eligibility of Joint Development Improvements Under Federal Transit Law." *Federal Register.* Vol. 72, No. 25. February 7, 2007. http://www.gpo.gov/fdsys/pkg/FR-2007-02-07/html/E7-1977.htm.

[19] Ibid.

[20] FTA. "Joint Development Frequently Asked Questions." http://www.fta.dot.gov/about_FTA_11011.html. Accessed July 26, 2012.

[21] Ibid.

STEPS/TIMING

Putting together a joint development project is a complex and multiagency effort. The process for determining if proceeds from the sale of the 19-acre parcel can be applied to a TOD infrastructure project on the 29-acre parcel could begin immediately through discussions with regional FTA staff. In the short term, UTA could also conduct a pre-feasibility analysis in house or hire advisors in real estate and/or finance to assess the potential revenue that could be generated under current or future real estate market conditions.

SHARED PARKING

In addition to developing a strategy to pay for additional parking in the station area, UTA and Sandy City could consider a strategy for shared parking that could reduce the need for costly structured parking. The Sandy Civic Center Station is near several major facilities that already provide a significant number of parking spaces. As development occurs in the station area, the actual need for parking in the station area and potential to use other existing parking in the area could be reassessed. A shared parking strategy could reduce the need for additional parking spaces in the station area, thereby reducing the overall costs of the project and helping to make TOD more feasible.

STEPS/TIMING

UTA and Sandy City can begin an evaluation of the potential for shared parking as parking needs are defined.

NEXT STEPS

Because UTA owns the property in question, it would be the lead entity in this effort. However, UTA would need to work with FTA, Sandy City, and potential developers.

- **Step 1: Revisit the Master Plan.** UTA could consider whether it needs all 48 acres of land that it currently owns in the area. If the UTA agrees that the 19-acre parcel can be severed from the 29-acre site, then UTA could pursue a disposition strategy for the 19-acre parcel. Selling the 19-acre parcel would provide revenue to apply towards TOD infrastructure needed to support development in the 29-acre site. UTA could pursue a deed restriction on the sale of the parcel that requires a level of density that is compatible with Sandy City's and UTA's TOD goals for the area. UTA would need to work with the selected developer of the site to ensure that the development program for the remaining 29-acre parcel is feasible.

- **Step 2: Work with FTA on Joint Development.** UTA could contact regional FTA staff to understand the potential to use proceeds from the sale of the 19-acre site on TOD infrastructure on the 29-acre site.

D. WHEAT RIDGE, COLORADO

LOCAL CONTEXT

Wheat Ridge is a small, inner-ring suburban city located at the end of the planned Gold Line Corridor (Exhibit D-7). The Gold Line is an 11.2-mile commuter rail line planned to originate from Denver Union Station and serve seven stations as it passes through northwest Denver, Adams County, and the city of Arvada, terminating at the Ward Road Station in Wheat Ridge. The Gold Line is part of the Regional Transportation District of Denver's (RTD) "FasTracks" program, a voter-approved plan to expand rail and bus service across the Denver region. The Gold Line is scheduled to open in mid-2016.

The Ward Road Station area in Wheat Ridge is currently characterized by industrial uses and vacant lots, with a mix of residential, commercial, and industrial uses surrounding the site (Exhibit D-8).

TOD PROJECTS AND TOD INFRASTRUCTURE NEEDS

STATUS OF TOD PROJECTS

The city of Wheat Ridge and RTD have entered into an intergovernmental agreement that outlines the responsibilities for each agency. Under the agreement, RTD is required to build and/or fund access roads to the station area in addition to the station itself and required parking. The city has not yet identified the funding sources and financing mechanisms to pay for additional infrastructure improvements in the surrounding area.

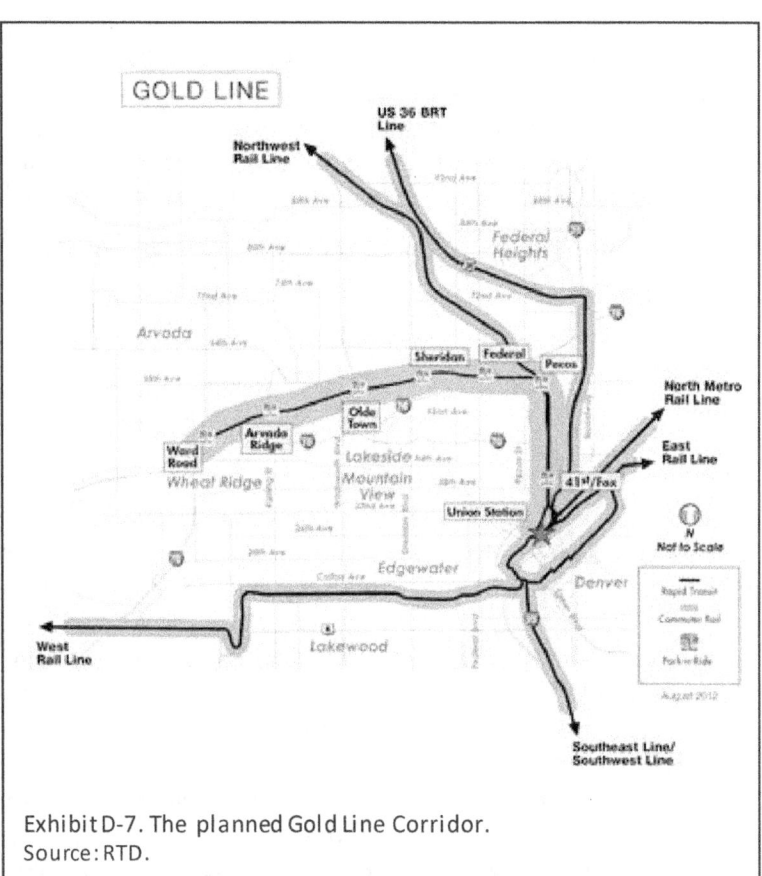

Exhibit D-7. The planned Gold Line Corridor.
Source: RTD.

While adjacent property owners generally support the long-term vision for the area, and at least one property owner has shown considerable interest in the potential for redevelopment, there are currently no planned developments. Property owners and potential developers might be waiting until the commuter rail line is operating to assess redevelopment opportunities. However, the city is concerned that if the infrastructure around the station is built in a low-density, suburban pattern at the outset, before private participation can be leveraged, it might not be possible to build a higher-density, mixed-use development.

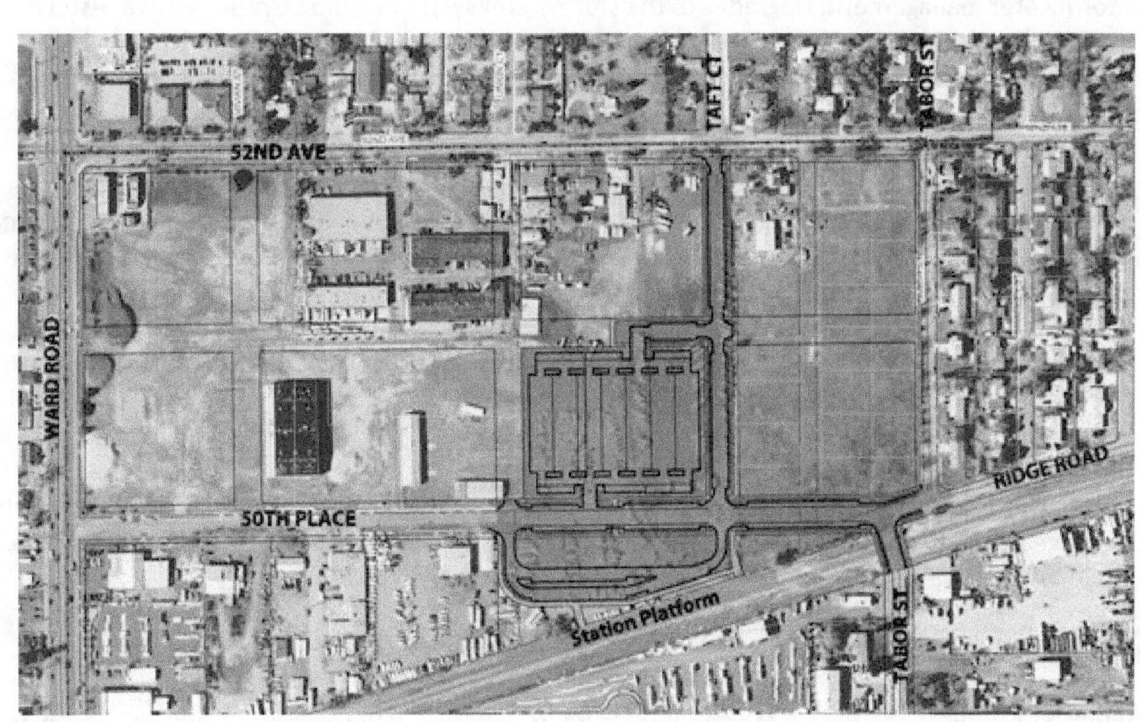

Exhibit D-8. Ward Road Station platform location and preliminary design. This figure shows the location for the station platform between 50th Place and the rail line near Taft Court. The areas shaded in green indicate the station platform, the associated parking area, and street improvements to be funded with the station. Planned future street improvements are indicated as grid lines.
Source: City of Wheat Ridge, Colorado.

The Denver Regional Council of Governments (DRCOG) won a HUD Sustainable Communities Challenge Grant in 2011. The $4.5-million grant is focused on three FasTracks lines, including the Gold Line, and the grant includes $500,000 for a catalytic project on each line that could lead to other transformational changes in the area. DRCOG has not yet set the criteria for the selection of the catalytic projects, but Wheat Ridge will consider pursuing that funding opportunity.

INFRASTRUCTURE NEEDS

The city has identified the following infrastructure needs in the station area:[22]

- **Sewer and water:** Upgrades to the existing service would be needed to support the desired development in the area. The city of Wheat Ridge does not provide sewer and water, and the area has several utility districts, which can sometimes complicate implementation efforts.

---

[22] Based on personal communication with Ken Johnstone, Community Development Director, City of Wheat Ridge, and Sarah Showalter, Planner, City of Wheat Ridge, by Sarah Graham and Dena Belzer, Strategic Economics, July 18, 2011, and January 4, 2012.

- **Stormwater management:** Upgrades to the stormwater system would be required as a result of the station. The city intends to move to a regional system of stormwater management rather than relying on site-specific stormwater facilities to allow more efficient use of land in the station area.

- **Street grid:** RTD has committed to building and/or funding the surface streets in the immediate vicinity of the station and parking area. The city and/or private property owners will need to complete the street grid as development occurs. RTD's street improvements include making changes to the Tabor Street intersection, extending Taft Street to the station, and enhancing access to the station area from the south side of the rail tracks via Tabor Street.

- **Structured parking:** RTD will acquire the land for parking and will either construct a surface parking lot, as the plans currently describe, or allocate funding equivalent to the cost of the required surface parking to a structured parking project.

- **Pedestrian bridge:** The bridge would cross over the rail tracks from somewhere near or on 49th Place or Ridge Road on the south side of the tracks to the station platform. RTD has identified possible pedestrian bridge landing locations to enable future pedestrian access.

PRIORITIES FOR TOD INFRASTRUCTURE AND POTENTIAL FINANCING MECHANISMS

Although early indications were that providing structured parking in the area was a high priority in order to allow for more dense development within the station area, structured parking is not actually a barrier to development in the short term. Instead, the immediate constraints on development in the station area are the stormwater management needs and lack of street connectivity, particularly access to the station area from the south.

- **Stormwater:** Because stormwater management is an immediate barrier to development, and a solution to the issue could be a catalytic project for the area, it should be a first priority for the City of Wheat Ridge. Potential funding and financing tools include a new stormwater utility fee, calculated on the basis of impervious area, and a district-based financing tool, such as a metropolitan district or assessment district. As described in more detail in Appendix B, Section A-1, utility fees are charged for the use of public infrastructure or goods. Fees are typically set to cover a system's operating and capital expenses each year, which can include debt service for improvements to the system, or at least some portion of those expenses. The stormwater system is a potentially catalytic project in several respects: it could be eligible for funding as a catalytic project under DRCOG's HUD grant; it could present an opportunity for collaboration with the neighboring jurisdiction of Arvada, paving the way for a long-term partnership on infrastructure investments; and it could bring property owners together around a problem that affects not only potential development but also existing uses.

- **Street grid:** The roadways, sidewalks, and pedestrian bridge should be a relatively high priority but do not need to happen all at one time. Streets and sidewalks will be required as development occurs. The pedestrian bridge could be constructed early in the process, ideally concurrently with the construction of the station platform.

- **Sewer and water:** Although the city does not provide the infrastructure for sewer and water, it does own the streets where the utilities are located, and the sequencing of development could be affected by utility provision—in other words, development cannot occur if the utilities are not built.

The city could work with the utility districts on the sequencing of infrastructure projects, determining what utility infrastructure costs are reasonable for developers to pay. A value capture tool, such as TIF, an assessment district, or development impact fees, could be used to provide sewer and water in the station area.

- **Structured parking:** Although structured parking might allow for more compact development in the station area eventually, building it later might be more financially feasible. The Ward Road station will be an end-of-the-line station, and therefore it has a significant parking requirement. While RTD has committed to contributing funding equivalent to the cost of the required surface parking to a structured parking project, that amount might not defray enough of the structured parking costs to make this project an immediate priority. Another RTD rail line, the West Rail Line, will have an end-of-the-line station at the Jefferson County Government Center that could serve some of the same commuters as the Ward Road station. If ridership levels are not as high as anticipated at Ward Road, or if riders access the station by walking or using transit, a lower parking requirement might be warranted, reducing the costs of structured parking. Once the line is built and service is established, the amount of parking needed at the station will be clearer. A lower parking requirement could make structured parking more financially feasible.[23] Although charging for most parking is currently not permissible at transit stations in Colorado, implementing a fee for parking at the Ward Road station could help defray the cost of structured parking. That strategy would only work, however, with a corridorwide or regional approach to parking fees.[24] Other options for financing structured parking include value capture, TIF, and assessment districts.

- **Pedestrian Bridge:** As with the street grid, the pedestrian bridge will make walking more appealing and convenient and therefore could have a significant impact on property values. Therefore, the city could include the pedestrian bridge in a value capture strategy using tools such as TIF, an assessment district, or development impact fees. The city could also pursue grant funding for the pedestrian bridge.

TOD Infrastructure	Value Capture Tools			User / Utility Fees	Debt Tools	Federal Grants/ Regional Funds
	Assessment Districts	TIF	Developer Impact Fees			
Stormwater	X	X	X		X	X
Street grid	X	X		X		X
Sewer/water	X	X	X	X	X	
Structured parking	X	X	X		X	X
Pedestrian bridge	X	X				X

Exhibit D-8. Summary of TOD infrastructure financing options for Wheat Ridge.

---

[23] The Federal Transit Administration issued a record of decision approving the RTD Gold Line project in November 2009. The record of decision was based on a final environmental impact statement that assumed a certain number of parking spaces would be available at the Ward Road station on opening day in 2015, with additional spaces added by 2030. A reduction in the parking requirement at the Ward Road station would require revision of the environmental impact statement and record of decision.

[24] More details about corridor-wide approaches to parking are in Chapter IV, Section B.

NEXT STEPS

- **Step 1: Create and strengthen partnerships.** Although the city could lead the implementation process, establishing and strengthening partnerships with key stakeholders will be important. The city could create partnerships that allow it to access different funding sources. By working collaboratively with neighboring jurisdictions and service districts, RTD, developers, and property owners, Wheat Ridge would be better positioned to implement further steps. The city of Arvada, in particular, could be an important partner in taking a regional approach to infrastructure projects such as parking and stormwater. The city and RTD have already codified a working relationship through the intergovernmental agreement regarding the station area.

- **Step 2: Have a clear plan.** It is critical for Wheat Ridge to examine its infrastructure needs and set clear priorities for infrastructure investments. Although Wheat Ridge has already prepared the Northwest Sub-area Plan, that planning document does not include an infrastructure project list or detailed implementation strategy. An implementation strategy could include a funding and financing section that examines the "creation of a district or districts for the subarea to provide a mechanism to finance, construct and maintain parking facilities, drainage facilities, parks and recreation facilities and streetscape improvements" as recommended in the Northwest Sub-area Plan.[25] A detailed implementation strategy financing plan could prioritize the components of the plan and break them down into phases matched with funding sources and financing mechanisms.

  Because resources are scarce, it is critical to set priorities and determine where best to apply limited funds. Through the implementation strategy, the city could set implementation priorities based on overcoming barriers to development, funding availability, and market strengths, not just on solving the "biggest" problems with the highest price tags.

- **Step 3: Leverage Opportunities.** The HUD Sustainable Communities Challenge Grant administered by DRCOG offers Wheat Ridge the opportunity to pursue a regional approach to its stormwater management needs. As described above, the grant total includes $500,000 for a catalytic project on each line, and Wheat Ridge could pursue that funding opportunity. This project could provide a spark and lay the foundation for other development in the area.

---

[25] City of Wheat Ridge. *Northwest Subarea Plan.* 2006.
http://www.ci.wheatridge.co.us/DocumentCenter/Home/View/565.